Hodder Gibson

Scottish Examination Materials

Rebecca Harrison
4K

INTERMEDIATE 2

CHEMISTRY

Revision Notes & Questions

Norman Conquest
with
George Thomson

Hodder Gibson

A MEMBER OF THE HODDER HEADLINE GROUP

The Publishers would like to thank the following for permission to reproduce copyright material:

Acknowledgements
Extracts from question Papers are reprinted by permission of the Scottish Qualifications Authority.

Every effort has been made to trace all copyright holders, but if any have been inadvertently overlooked the Publishers will be pleased to make the necessary arrangements at the first opportunity.

Although every effort has been made to ensure that website addresses are correct at time of going to press, Hodder Gibson cannot be held responsible for the content of any website mentioned in this book. It is sometimes possible to find a relocated web page by typing in the address of the home page for a website in the URL window of your browser.

Hodder Headline's policy is to use papers that are natural, renewable and recyclable products and made from wood grown in sustainable forests. The logging and manufacturing processes are expected to conform to the environmental regulations of the country of origin.

Orders: please contact Bookpoint Ltd, 130 Milton Park, Abingdon, Oxon OX14 4SB. Telephone: (44) 01235 827720. Fax: (44) 01235 400454. Lines are open 9.00 – 5.00, Monday to Saturday, with a 24-hour message answering service. Visit our website at www.hoddereducation.co.uk. Hodder Gibson can be contacted direct on: Tel: 0141 848 1609; Fax: 0141 889 6315; email: hoddergibson@hodder.co.uk

© Norman Conquest 2007
First published in 2007 by
Hodder Gibson, an imprint of Hodder Education,
a member of the Hodder Headline Group
2a Christie Street
Paisley PA1 1NB

Impression number   5 4 3 2 1
Year   2010 2009 2008 2007

Cover photo © Bernard Edmaier/Science Photo Library
Illustrations by Fakenham Photosetting Limited
Typeset in Garamond BE 10.5/12pt by Fakenham Photosetting Limited, Fakenham, Norfolk
Printed and bound in Great Britain by Martins the Printers, Berwick-upon-Tweed

A catalogue record for this title is available from the British Library

ISBN-10: 0-340-92706-2
ISBN-13: 978-0-340-92706-9

Contents

iii

Contents

Introduction

This book is a useful revision aid both for normal classroom use and for the individual student working alone. The Intermediate 2 Chemistry course consists of three units, and revision notes covering essential theory are given for each of these, along with questions dealing with knowledge and understanding as well as problem solving. Answers are provided for all questions so that the student receives instant feedback and can easily be directed back to areas which need more attention. *Prescribed Practical Activities*, which provide step-by-step guides to practical investigations, also appear throughout the book.

The questions in the first three sections of the book, which correspond directly to the three units of the Intermediate 2 Chemistry course, reflect the more straightforward nature of the questions that will be met in the end-of-unit tests. More challenging questions, of the sort that will be used to grade student awards, are given in Section 4 along with revision of material which cuts across the course or is found at various points in it. Section 4 will be of considerable use when the student is preparing for the external examination.

The integrated nature of this book, providing revision notes, questions at different levels of difficulty and answers to all questions, makes it a particularly useful revision aid, both prior to end-of-unit tests and before the external examination at the end of the course.

Section 1

BUILDING BLOCKS

1.1 Substances

Elements

○ Everything in the world is made from about 100 **elements**.

○ There is a different *symbol* for every element – for example C for carbon.

○ Chemists have classified elements by arranging them into a **Periodic Table**.

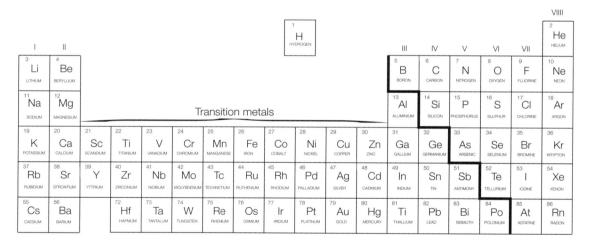

○ The majority of the first 92 elements in the Periodic Table occur naturally (mostly in compounds) – the rest have been made by scientists.

○ Most elements are solids, a few are gases and two – bromine and mercury – are liquids (at room temperature).

○ Elements can also be classified as being either **metals** or **non-metals** – most are metals.

○ A column of elements in the Periodic Table is called a **group**. A row of elements is called a **period**.

○ Elements in the same group have similar chemical properties.

○ Important groups include:

– Group 1 – the **alkali metals** (reactive)

1

– Group 7 – the **halogens** (reactive non-metals)

– Group 0 – the **noble gases** (very unreactive)

○ The *transition metals* are an important block of elements between Groups 2 and 3.

Compounds

○ **Compounds** are formed when elements react together.

○ Compounds with the ending *-ide* usually contain *two* elements.

○ Compounds with the ending *-ite* or *-ate* usually contain *three* elements, one of which is *oxygen*.

○ Compounds tend not to resemble the elements from which they are made.

○ To separate the elements in a compound requires a chemical reaction.

Mixtures and separation techniques

○ A **mixture** is formed when two or more substances mingle together without reacting.

○ *Air* is a mixture of gases – dry air contains 78% nitrogen, 21% oxygen, about 1% noble gases (mainly argon) and 0.03% carbon dioxide.

○ Separating the substances in a mixture does *not* involve a chemical reaction.

○ Separation techniques include:

– **filtration** to separate an insoluble solid from a liquid (see Figure 1)

– **evaporation** to separate a dissolved solid from a liquid (see Figure 2)

– **distillation** to separate a mixture of liquids (see Figure 3)

– paper **chromatography** to separate a mixture of dissolved solids (see Figure 4).

Filtration

○ Insoluble solids can be separated from liquids by passage through filter paper.

○ The solid which remains behind in the filter paper is called the **residue.**

○ The liquid which passes through the filter paper is called the **filtrate.**

○ Precipitates can be separated from aqueous solutions of other substances using this technique.

Evaporation

○ A solid **solute** can be recovered from a **solution** by evaporating the **solvent**.

○ Soluble salts can be recovered from solutions using this technique.

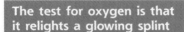

> **The test for oxygen is that it relights a glowing splint**
>
> Air does not give this result because it does not contain enough oxygen.

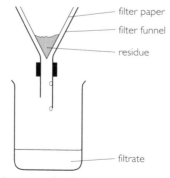

Figure 1 Filtration

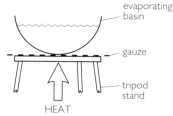

Figure 2 Evaporation

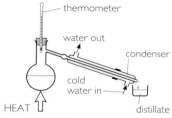

Figure 3 Distillation

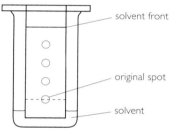

Figure 4 Chromatography

❍ If well-shaped crystals are required then the final crystallisation stage has to be carried out slowly.

Distillation

❍ A mixture of liquids can be separated by distillation – the liquid with the lowest boiling point distils off first.

❍ Ethanol (boiling point 78 °C) can be separated from water (boiling point 100 °C) after the fermentation of sugars such as glucose.

Chromatography

❍ Small quantities of soluble compounds can be separated using paper chromatography.

❍ The separation of coloured substances, such as food colours or coloured inks, is easily seen.

❍ A drop of the mixture is spotted onto chromatography paper (or filter paper) and the bottom of the paper is placed in a suitable solvent (water can often be used).

❍ If the substances used are colourless – for example sugars or amino acids – a locating agent can be used to show their presence.

Solvents, solutes and solutions

❍ A **solvent** is a liquid in which a substance dissolves – for example water.

❍ A **solute** is a substance that dissolves in a solvent – for example common salt, sodium chloride.

❍ A **solution** is the mixture formed when a solute dissolves in a solvent – for example sea water, which contains many dissolved salts.

❍ A substance which dissolves in a liquid is said to be *soluble* – for example sodium chloride is soluble in water.

❍ A substance which does not dissolve in a liquid is said to be *insoluble* – for example sand is insoluble in water.

❍ A **concentrated solution** is one in which there is a lot of solute compared with the amount of solvent.

❍ A **dilute solution** is one in which there is little solute compared with the amount of solvent.

❍ A **saturated solution** is one in which no more solute can be dissolved.

❍ A solution is *diluted* by adding more solvent.

❍ Water is the most common solvent.

❍ **State symbols** can be used to indicate the state of a substance:

(aq) = aqueous (dissolved in water), **(g)** = gas, **(l)** = liquid, **(s)** = solid.

Some everyday chemical reactions

Fuels burning
Iron rusting
Copper corroding
Wine fermenting
Dough rising
Bread toasting
Eggs frying
Cakes baking
Food digesting
Indigestion remedies acting

Chemical reactions

○ All **chemical reactions** involve the formation of one or more *new substances*.

○ In many chemical reactions there is a change in *appearance* – for example a colour change happens, a gas is evolved or a precipitate is formed.

○ In many chemical reactions there is a detectable *energy change*.

○ **Exothermic reactions** release energy to the surroundings causing an *increase* in temperature.

○ The products of an exothermic reaction have *less* chemical energy than the reactants.

○ **Endothermic reactions** take in energy from the surroundings causing a *decrease* in temperature.

○ The products of an endothermic reaction have *more* chemical energy than the reactants.

Gas production and collection

○ Several gases can be produced from a solid and a solution using the apparatus shown in Figure 5. Solid A is placed in the flask and solution B is added via the thistle funnel.

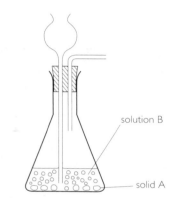

Figure 5

Gas produced	Solid A	Solution B
Hydrogen	Zinc	Dilute sulphuric acid
Carbon dioxide	Marble (CaCO$_3$)	Dilute hydrochloric acid
Oxygen	Manganese(IV) oxide	Hydrogen peroxide

Collection over water

○ If a gas is *not* very soluble in water then it may be collected over water as shown in Figure 6.

○ Many gases have a low solubility in water and can therefore be collected in this way.

○ Gases which can be collected over water include oxygen, nitrogen, carbon dioxide, the noble gases and all hydrocarbon gases.

○ Gases which are appreciably soluble in water, and therefore *cannot* be collected in this way, include ammonia, sulphur dioxide and nitrogen dioxide.

○ Collection over water does not provide a *dry* gas.

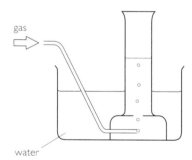

gas

water

Figure 6

Collection by downward displacement of air

○ Gases which are *less* dense than air may be collected by downward displacement of air as shown in Figure 7.

○ Gases which can be collected in this way include hydrogen, helium, ammonia and methane.

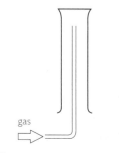

gas

Figure 7

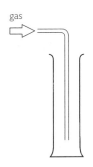

○ Gases which are less dense than air have a formula mass of less than 29.
○ Collection by this method provides a *dry* sample of gas.

Collection by upward displacement of air

○ Gases which are *more* dense than air may be collected by upward displacement of air as shown in Figure 8.
○ Gases which can be collected in this way include carbon dioxide, sulphur dioxide and nitrogen dioxide.
○ Gases which are more dense than air have a formula mass greater than 29.
○ Collection by this method provides a *dry* sample of gas.

Figure 8

QUESTIONS

1 Which of the following elements occurs naturally uncombined?
   A Fluorine
   B Nitrogen
   C Magnesium
   D Potassium

2 When potassium nitrate dissolves in water, the process is slightly endothermic. If the initial water temperature is 20 °C, what is the most likely final temperature on dissolving 1 g of potassium nitrate in 20 cm³ of water?
   A −17 °C
   B 17 °C
   C 23 °C
   D 30 °C

3 The solubility of a certain substance changes in a steady manner with rise in temperature.

Temperature /°C	Mass of substance dissolved in 100 g of water/°C
50	40
75	90

The most likely mass of substance that will dissolve in 100 g of water at 40 °C is
   A 30 g
   B 40 g
   C 50 g
   D 60 g

4 Soda water is made by dissolving carbon dioxide in water. Which of the following correctly describes these substances?

	Carbon dioxide	Water	Soda water
A	Solvent	Solute	Solution
B	Solute	Solvent	Solution
C	Solvent	Solution	Solute
D	Solute	Solution	Solvent

5 Which of the following is not an alkali metal?
   A Barium
   B Lithium
   C Rubidium
   D Caesium

6 The gas which relights a glowing splint is
   A hydrogen
   B carbon dioxide
   C nitrogen
   D oxygen

7 The diagram shows a part of the Periodic Table. The letters used do not represent chemical symbols.

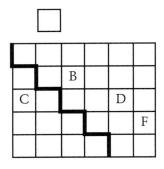

    a) Identify the noble gas.
    b) Identify the halogen.
    c) Identify the two elements in the same group.
    d) Identify the element in Group 5.

8 Over 100 elements are known. Six are shown below.

    Krypton   **Bromine**   **Oxygen**

    Carbon   **Iron**   **Calcium**

    a) Identify the transition metal.
    b) Identify the element in Group 6.
    c) Identify the element that is very unreactive.
    d) Identify the liquid non-metal.

9 Copy and complete the following sentences.
    Chemists have classified elements by arranging them in the _____ _____.
    Most of the first 92 elements occur _____ (mostly in compounds).
    A column of elements in the Periodic Table is called a _____.
    Elements in the same group have similar _____ _____.

10 The first 20 elements (hydrogen to calcium) in the Periodic Table were all discovered before the twentieth century began. Use the data booklet to help you answer the following questions about these elements.
    a) Which two elements were known in prehistoric times?
    b) Which elements were discovered in the eighteenth century?
    c) Which was the last of the first 20 elements to be discovered?

11 Use the data booklet to help you answer these questions.
    a) Which of the following are metals and which are non-metals:
    thallium, polonium, tellurium?
    b) Which of the following do not occur naturally either as the element or in a compound:
    protactinium, astatine, francium?
    c) Which of the alkali metals would be a liquid on a hot summer day when the temperature reached 30 °C?

12 Copy and complete the following tables.

a)

Compound	Elements present		
Hydrogen oxide	Hydrogen		Oxygen
Lead chloride	_____		_____
Sodium sulphide	_____		_____
Calcium _____	_____		Bromine

b)

Compound	Elements present		
Lithium sulphate	Lithium	Sulphur	Oxygen
Zinc carbonate	_____	_____	_____
_____ nitrate	Copper	_____	_____
_____ sulphite	Barium	_____	_____
Potassium nitrite	_____	_____	_____

13 A solution is a mixture formed when one or more solutes dissolves in a solvent. Give five examples of this type of mixture that might be found in the average home.

14 Lager and whisky are both mainly water containing dissolved alcohol. Lager contains about 5% alcohol, whereas whisky contains about 40% alcohol.
    A Solvent
    B Dilute solution
    C Concentrated solution
    D Solute
    a) Which term best describes the alcohol in these liquids?
    b) Which term best describes whisky?

15 Which of these involves a chemical reaction?
    A Water evaporating
    B Apples rotting
    C Fat melting
    D Electric light bulbs glowing

16 If a piece of aluminium foil is held under mercury and scratched with a sharp object, something unusual happens to it when it is removed. Almost immediately a white substance appears to grow up from the scratch marks. As this happens, the foil becomes distinctly warm. What evidence is there that a chemical reaction has taken place?

(1.2) Reaction rates

Factors affecting rate of reaction

○ Chemical reactions are speeded up by:
- decreasing particle size
- increasing concentration
- increasing temperature.

Examples

○ *Particle size*: small marble chips react faster with acid than large ones; sticks burn faster than a log.
○ *Concentration*: marble chips react faster with concentrated acid than with dilute acid; concentrated bleach acts faster on a stain than dilute bleach.
○ *Temperature*: an acid reacts faster with marble chips at a higher temperature; milk goes sour faster at room temperature than in a refrigerator.

Collision theory

○ The *collision theory* states that for a reaction to take place, particles must collide.
○ Not all collisions are successful and result in a reaction taking place – for example natural gas and oxygen do not react at room temperature.
○ Increasing the concentration of a reactant increases the collision rate and thus the number of successful collisions.
○ Decreasing the particle size of a solid reactant increases the surface area, which increases the collision rate and thus the number of successful collisions.

Following the course of a reaction

○ Reactions can be followed by measuring changes in concentration, mass and volume of reactants and products. Usually one of the following is monitored:
- the decrease in concentration of a reactant
- the increase in concentration of a product
- the volume of a gas given off
- the mass of a gas given off.
○ The average rate of a chemical reaction over a certain period of time is given by:

$$\text{average } \textbf{rate of reaction} = \frac{\text{change in quantity measured}}{\text{time taken}}$$

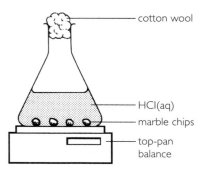

cotton wool

HCl(aq)

marble chips

top-pan balance

Figure 1

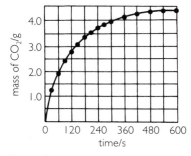

Figure 2

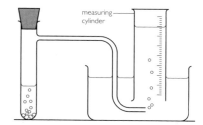

measuring cylinder

○ The *units* of reaction rate include the following:
  – $mol\,l^{-1}\,s^{-1}$ when a concentration is monitored
  – $cm^3\,s^{-1}$ when the volume of a gas evolved is monitored
  – $g\,s^{-1}$ when the mass of a gas evolved is monitored.

○ The reaction between a carbonate and an acid, such as that between marble and hydrochloric acid, can be studied using the apparatus shown in Figure 1. The mass of carbon dioxide given off can be recorded at appropriate time intervals.

○ Figure 2 shows a typical graph where the marble was in excess. From the graph the average rate of reaction over certain time intervals can be calculated:

In the first 60 seconds the average rate of reaction is given by

$$\text{average rate} = \frac{\text{mass of } CO_2 \text{ evolved}}{\text{time taken}}$$

$$= \frac{2.0\,g}{60\,s}$$

$$= 0.0333\,g\,s^{-1}$$

In the next 60 seconds the average rate of reaction is

$$\text{average rate} = \frac{0.8\,g}{60\,s}$$

$$= 0.0133\,g\,s^{-1}$$

○ The rate of reaction decreases because there is a decrease in acid concentration and a decrease in the surface area of the marble chips, as both are used up.

○ A graph of volume of gas evolved against time may be obtained if the reaction flask is attached to a gas syringe, or if the gas is collected over water in an upturned measuring cylinder.

○ If the rate of a reaction is known, then the time taken for a given amount of product to form or reactant to be used up can be calculated. For example, in the reaction of a carbonate with an acid if the rate of production of CO_2 is $4\,cm^3\,s^{-1}$ (4 centimetres cubed of carbon dioxide are formed every second) it is possible to calculate the time taken to produce $100\,cm^3$ of CO_2.

$$\text{rate} = \frac{\text{volume of } CO_2 \text{ produced}}{\text{time taken}}$$

$$\text{time taken} = \frac{\text{volume of } CO_2 \text{ produced}}{\text{rate}}$$

$$= \frac{100\,cm^3}{4\,cm^3\,s^{-1}}$$

$$= 25\,s$$

Measuring the rate of the iodide/persulphate reaction

○ Iodine is produced in the reaction between solutions of potassium iodide and sodium persulphate. However, if thiosulphate ions are also present, no blue colour with starch appears until all of the thiosulphate ions have reacted with the iodine being produced. It is not possible to measure the mass of iodine produced before the appearance of the blue colour, but an experiment can be devised in which the time is noted for a constant mass of iodine to be formed.

$$\text{Rate of reaction} = \frac{\text{mass of iodine produced}}{\text{time taken}} = \frac{\text{a constant}}{\text{time taken}} = \frac{k}{t}$$

$$\text{Rate of reaction} \propto \frac{1}{t}$$

○ For all reactions, the rate of reaction, or of a stage in a reaction, is proportional to the reciprocal of the time taken. If the rate of reaction is high then the time taken will be small; but if the rate of reaction is low the time taken will be large.

○ In the case of the reaction between solutions of potassium iodide and sodium persulphate, a graph of $1/t$ (a measure of reaction rate) against sodium persulphate concentration is a straight line. This indicates that reaction rate is *directly* proportional to sodium persulphate concentration.

Prescribed Practical Activity

Effect of concentration on reaction rate

• The aim of this experiment is to find the effect of varying the concentration of sodium persulphate solution on the rate of its reaction with potassium iodide solution.

• Using syringes, measure $10\,cm^3$ of sodium persulphate solution and $1\,cm^3$ of starch solution into a small dry beaker placed on a white tile or piece of white paper.

• Using a syringe, measure $10\,cm^3$ of potassium iodide solution containing sodium thiosulphate solution into the beaker and start timing.

• Note the time taken for the mixture to suddenly go dark blue.

• Wash and dry the beaker.

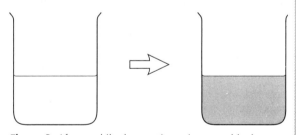

Figure 3 After a while the reaction mixture suddenly becomes dark blue

• Repeat the procedure, but this time dilute the persulphate solution by using syringes to measure $8\,cm^3$ of sodium persulphate solution, $2\,cm^3$ of deionised water and $1\,cm^3$ of starch solution into the beaker.

• Repeat using $6\,cm^3$ of sodium persulphate solution, $4\,cm^3$ of deionised water and $1\,cm^3$ of starch solution.

continued ➤

- Repeat using 4 cm³ of sodium persulphate solution, 6 cm³ of deionised water and 1 cm³ of starch solution.
- A typical set of results is shown in the table below.
- The graph of reaction rate/s^{-1} against volume of sodium persulphate/cm³ is shown in Figure 4. This is a straight line indicating direct proportionality between reaction rate and concentration of sodium persulphate solution. This is because, in this case, the volume of persulphate used is also a measure of its concentration.
- The volumes and concentrations need not be those given in the table, but the following *must* be kept constant:
  - the total volume of the mixture
  - the volume and concentration of the potassium iodide solution containing the sodium thiosulphate solution
  - the temperature of the mixture.

- The only **variable** should be the volume of the sodium persulphate solution – which directly affects its concentration in the reaction mixture.

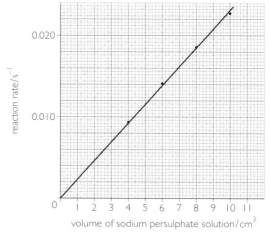

Figure 4

Experiment	1	2	3	4
Volume of sodium persulphate (aq)/cm³	10	8	6	4
Volume of deionised water/cm³	0	2	4	6
Time for dark blue colour to appear/s	44	54	71	107
Reaction rate/s^{-1}	0.0227	0.0185	0.0141	0.0093

Measuring the rate of the acid/thiosulphate reaction

○ In the reaction between solutions of hydrochloric acid and sodium thiosulphate, a suspension of finely divided sulphur is produced which eventually makes the mixture opaque. It is not possible to measure the mass of sulphur produced in a given time, but an experiment can be devised in which the time is noted for a constant mass of sulphur to be formed.

$$\text{Rate of reaction} = \frac{\text{mass of sulphur produced}}{\text{time taken}} = \frac{\text{a constant}}{\text{time taken}} = \frac{k}{t}$$

$$\text{Rate of reaction} \propto \frac{1}{t}$$

Prescribed Practical Activity

Effect of temperature on reaction rate

Safety note

The sulphur dioxide which is formed during the experiment is toxic and can cause an asthmatic attack

- The aim of this experiment is to find the effect of varying the temperature on the rate of reaction between sodium thiosulphate solution and hydrochloric acid.
- Pour about 100 cm³ of sodium thiosulphate solution into a medium-sized glass beaker.
- Lightly draw a cross on a piece of paper.
- Using a syringe, measure 20 cm³ of the sodium thiosulphate solution into a small beaker which is placed on the cross.

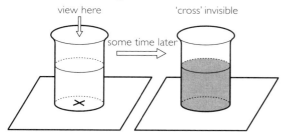

Figure 5

- Using a syringe, measure 1 cm³ of hydrochloric acid into the small beaker, stir and start timing.
- View the reaction mixture from above and note the time taken for the cross to disappear.
- Note the temperature of the reaction mixture.
- Wash and dry the small beaker.
- Heat the remaining sodium thiosulphate solution to about 30 °C.
- As before, measure 20 cm³ of sodium thiosulphate solution into the small beaker, place it on the cross and add 1 cm³ of hydrochloric acid. Stir and note the time taken for the cross to disappear. Note the temperature of the mixture.

- Heat the remaining sodium thiosulphate solution to about 40 °C and repeat the procedure; then to about 50 °C and again repeat the procedure. Do not exceed 50 °C.
- A typical set of results is shown in the table below.

Temperature/°C	Time/s	Reaction rate/s⁻¹
20	80	0.0125
32	40	0.0250
39	27	0.0370
48	16	0.0625

- As expected, the rate of reaction increases with rise in temperature, but as can be seen in Figure 6, the graph of reaction rate/s⁻¹ against temperature/°C is a curve, not a straight line.
- In this experiment, the reaction rate doubles if there is a rise of about 12 °C.
- The volumes, concentrations and temperatures need not be those given here, but the following *must* be kept constant:
  – the volume and concentration of the hydrochloric acid
  – the volume and concentration of the sodium thiosulphate solution.
- Ideally the same clean beaker should be used, and the same person should observe the clouding over of the cross.
- The only variable should be the temperature of the mixture.

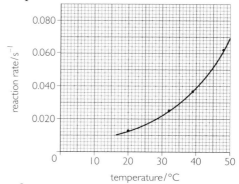

Figure 6

11

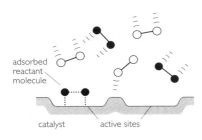

adsorbed
reactant
molecule

catalyst   active sites

Figure 7

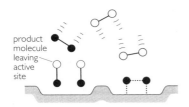

product
molecule
leaving
active
site

Figure 8

Catalysts

○ **Catalysts** are substances which speed up some reactions, but are not used up and can be recovered unchanged at the end of the reaction.

○ A **homogeneous catalyst** is one which is in the same state as the reactants. For example, pink cobalt(II) ions in aqueous solution catalyse a reaction between hydrogen peroxide and tartrate ions, which are also in aqueous solution. During the course of this reaction green cobalt(III) ions are formed, but pink cobalt(II) ions are reformed at the end of it.

○ A **heterogeneous catalyst** is one which is in a different state from the reactants – for example manganese(IV) oxide, which is a solid, catalyses the decomposition of hydrogen peroxide solution.

○ Heterogeneous catalysts are usually solids and work by the **adsorption** of reactant molecules onto active sites (see Figure 7). Reactant molecules form bonds with the catalyst.

○ The product molecules then leave the catalyst and the active site used becomes available for the adsorption of another reactant molecule (see Figure 8).

○ Rhodium, platinum and palladium act as heterogeneous catalysts in the **catalytic converters** fitted in car exhaust systems. Among the reactions that they catalyse, when hot, are:

carbon monoxide + nitrogen oxides → carbon dioxide + nitrogen
unburned hydrocarbons + oxygen → carbon dioxide + water.

○ The surface activity of a catalyst can be reduced by **catalytic poison**. This occurs when unwanted particles are preferentially adsorbed onto the surface of a catalyst – thus reducing the number of active sites available. For example, lead compounds in leaded petrol poison the transition metal catalysts in a catalytic converter. Cars fitted with these must use unleaded petrol.

○ Impurities in reactants result in industrial catalysts having to be regenerated or renewed.

○ There are many everyday examples of the use of catalysts – for example platinum is found in some heated hair-styling brushes where it catalyses the reaction between the hydrocarbon fuel inside the handle and oxygen in the air.

○ Catalysts are widely used in industrial processes – for example aluminium oxide is used as a catalyst in the cracking of hydrocarbons in the oil industry. During the reaction the aluminium oxide becomes coated with carbon – it is regenerated by burning this off.

○ **Enzymes** are biological catalysts. They catalyse the chemical reactions which take place in the living cells of plants and animals – for example the enzyme invertase catalyses the reaction of sucrose and water to give glucose and fructose.

○ Enzymes are used in the manufacture of cheese, yoghurt, bread, wine, beer, lager, whisky and biological detergents. There are many everyday examples of enzymes and they are used in many industrial processes.

QUESTIONS

1 Many transition metals can act as catalysts. Which of the following is a transition metal?
   A Barium
   B Gallium
   C Vanadium
   D Antimony

2 Kurt intended to study the effect of changing the acid concentration on the reaction between nitric acid and limestone to produce carbon dioxide gas. He decided to monitor the mass of the flask as the reaction proceeded. Which of the following need *not* be kept constant?
   A The temperature of the acid
   B The shape of the flask
   C The particle size of the limestone
   D The mass of the limestone

3 Copy and complete the following sentence: Chemical reactions are *slowed* down by _____ particle size, _____ concentration and _____ temperature.

4 Use the collision theory to explain why:
   a) increasing the concentration of a reactant increases reaction rate,
   b) decreasing the particle size of a solid reactant increases reaction rate.

5 a) Explain the meaning of the word *catalyst*.
   b) Distinguish between the terms *homogeneous* catalyst and *heterogeneous* catalyst.
   c) Rhodium, platinum and palladium are present as catalysts in the catalytic converters found in motor car exhaust systems.
   i) To what type of catalysts do they belong?
   ii) What are carbon monoxide and oxides of nitrogen changed into when they pass through the hot catalyst chamber?
   iii) Explain why petrol containing lead compounds must not be used in cars which have catalytic convertors.

6 The enzyme catalase dissolves in water and the solution catalyses the decomposition of a solution of hydrogen peroxide in water.
   a) What type of catalysis is taking place?
   b) What is the role of enzymes in plants and animals?
   c) Name two products, sold commercially, which are made with the help of enzymes.

7 Richard added 3 g of marble chips to 10 cm³ of dilute hydrochloric acid. He noted the volume of carbon dioxide produced every 10 seconds until the reaction was over. The apparatus used and the graph obtained were as follows.

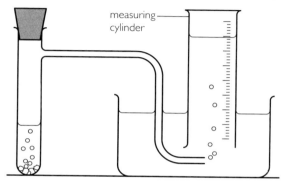

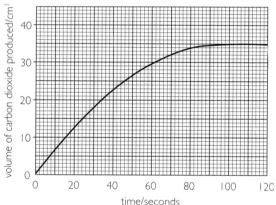

   a) Calculate the average rate of reaction during *each* of the first *three* 20 second intervals of the reaction.
   b) Give one possible reason for the slowing down of the reaction.
   c) Zeenat repeated the experiment, but used 3 g of even smaller chips than Richard. Sketch the graph and add a dotted line to show the type of graph that Zeenat would be expected to obtain.
   d) From Zeenat's experiment it was found that the rate of formation of CO_2 at the start of the reaction was 2 cm³ s⁻¹. Assuming the rate was constant, calculate the time required to produce 18 cm³ of CO_2.

8 The reaction between sodium thiosulphate solution and hydrochloric acid can be used to investigate the effect of temperature on reaction rate. The most noticeable change is the production of pale yellow sulphur in the reaction mixture, but it is not possible to find out the mass of sulphur produced at any given time. Explain how the reaction is carried out so that the time for a given, fixed mass of sulphur to be formed is obtained.

9 An experiment was set up to find out how quickly a compound called DIMP was hydrolysed when heated in acidic solution. The graph shows how the concentration of DIMP changed during the reaction.

a) By how much did the concentration of DIMP fall in the first 400 s?

b) Calculate the average rate of reaction, in $mol\,l^{-1}\,s^{-1}$, between 0 and 400 s.

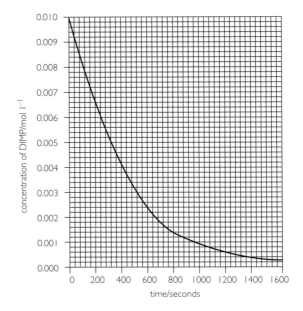

1.3 The structure of the atom

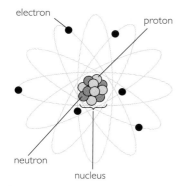

electron · proton · neutron · nucleus

Figure 1

Atoms

○ Every element is made up of very small particles called **atoms**.
○ All atoms have an extremely small positively charged central part called the **nucleus**.
○ The nucleus of every atom is positively charged due to the presence of **protons**, each of which has a single positive charge.
○ Negatively charged particles, called **electrons**, move around the outside of the nucleus.
○ Each electron has a single negative charge.
○ All atoms are electrically neutral because the numbers of protons and electrons are equal.
○ Electrons are arranged around each nucleus in special layers called **energy levels** or **shells.**
○ Elements with the same number of outer electrons have similar chemical properties.
○ Elements in the same group of the Periodic Table have the same number of outer electrons (and are therefore chemically similar).
○ Almost all atoms have **neutrons** in their nuclei – neutrons carry no charge.
○ Protons and neutrons are much heavier than electrons.

Particle	Charge	Relative mass
Proton	+1	1
Neutron	0	1
Electron	−1	0 (almost)

Atomic number, mass number and isotopes

○ The number of protons in the atoms of a particular element is fixed.
○ The number of neutrons in the atoms of an element can vary.
○ Most elements are made up of more than one kind of atom.
○ The **atomic number** of an atom is the number of protons in its nucleus.
○ Elements in the Periodic Table are arranged in order of increasing atomic number.
○ The **mass number** of an atom is the total number of protons and neutrons in its nucleus.
○ The number of neutrons in an atom is equal to the mass number minus the atomic number.

15

○ **Isotopes** are atoms of the same element which have different numbers of neutrons – they have the same atomic number but different mass numbers.

○ For any isotope, a special symbol can be written to show its mass number and atomic number – for example the isotope of chlorine with mass number 35 and atomic number 17 can be written:

$$\text{mass number} \longrightarrow ^{35}_{17}\text{Cl}$$
$$\text{atomic number} \longrightarrow$$

○ Most elements exist as a mixture of isotopes – for example chlorine consists of two isotopes, one with mass number 35 and one with mass number 37.

Relative atomic mass

○ The **relative atomic mass** (**RAM**) of an element is the average of the mass numbers of its isotopes, taking into account the proportions of each.

○ The relative atomic mass of an element is rarely a whole number because of the existence of isotopes – for example chlorine has two naturally occurring isotopes, ^{35}Cl (75%) and ^{37}Cl (25%); it has an RAM of 35.5.

○ The RAMs quoted in data books are often given to the nearest whole number – for example the RAM of magnesium is 24.3, but is usually given as 24.

QUESTIONS

1 An atom containing nine protons, ten neutrons and nine electrons will have
A the symbol $^{18}_{10}\text{Ne}$
B an atomic number of 10
C a mass number of 18
D seven outer electrons

2 An electrically neutral atom possesses 46 neutrons and 36 electrons.
a) The atomic number of the atom is
A 82   B 46   C 36   D 10
b) The symbol for the atom is
A $^{36}_{10}\text{Kr}$ B $^{82}_{36}\text{Kr}$ C $^{82}_{46}\text{Kr}$ D $^{46}_{36}\text{Kr}$
c) An atom of another isotope of the same element could possess
A 46 protons and 36 electrons
B 46 neutrons and 48 electrons
C 46 protons and 46 neutrons
D 48 neutrons and 36 electrons

3 Atoms of the isotopes $^{14}_{6}\text{X}$ and $^{14}_{7}\text{Q}$
A possess the same number of protons
B possess the same number of electrons
C possess the same number of neutrons
D belong to elements in the same row of the Periodic Table

4 An element with eight electrons in its outer energy level could have an atomic number of
A 54   B 50   C 37   D 35

5 Atoms X and Y have the same mass numbers, but the atomic number of X is greater than that of Y. It can therefore be said that:
A X and Y are isotopes of the same element
B atoms of X and Y contain the same number of protons
C atoms of Y contain more neutrons than atoms of X
D atoms of Y contain more electrons than atoms of X

6 An element has a relative atomic mass of 55.8. It may therefore be concluded that:

   A the element is made up of two isotopes

   B the element must contain at least one isotope with a mass number of 56 or greater

   C the most abundant isotope has a mass number of 56

   D one of the isotopes has a mass number of 55

7 Use this list to answer the questions that follow.

   A Number of protons in an atom

   B Number of neutrons in an atom

   C Number of electrons in an atom

   D Number of protons and neutrons in an atom

   E Number of outermost electrons in an atom

   F Group number in the Periodic Table

Which two options (A–F) contain numbers that:

  a) are both equal to the atomic number of an atom?

  b) when added together are equal to the mass number of an atom?

  c) are both equal to four for carbon atoms?

  d) change from one isotope of an element to another?

8 When measured accurately, it is found that very few elements have relative atomic masses that are whole numbers – for example copper has a relative atomic mass of 63.5. Give an explanation for this.

9 Boron consists of two isotopes with mass numbers 10 and 11.

  a) Explain the meaning of the word *isotope*.

  b) How many neutrons are present in each atom of the boron isotope with a mass number of 11?

  c) The accurate value for the relative atomic mass of boron is 10.8. Explain which isotope is present in the greatest amount.

10 Nickel has a relative atomic mass of 58.7, but no nickel atom has this mass number. Explain this apparent contradiction.

11 Use the information in the data booklet to find the missing terms in the table below.

Element	Symbol	Atomic number	Electron arrangement
Beryllium	Be	4	2, 2
(a)	Ge	(b)	(c)
(d)	(e)	38	(f)

12 With the help of the information in the data booklet, find the missing terms in the table below. Each line refers to a different atom.

Isotope	Mass number	Atomic number	Number of protons	Number of neutrons	Number of electrons
$^{3}_{1}H$	3	1	1	2	1
$^{14}_{6}C$	(a)	(b)	(c)	(d)	(e)
(f)	21	(g)	(h)	(i)	10
(j)	(k)	(l)	26	30	(m)

(1.4) Bonding, structure and properties

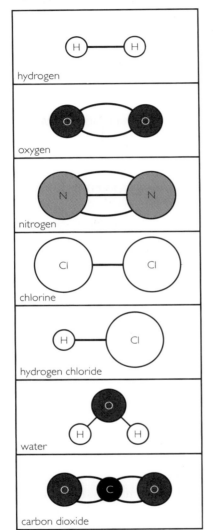

Figure 1 Molecular models

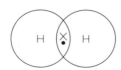

Figure 2 A hydrogen molecule

○ Only the noble gases exist as individual atoms not bonded to other atoms.

○ In all other substances, atoms are held together by *bonds*.

○ In forming bonds, the main group elements usually achieve the stable electron arrangement of a noble gas.

Covalent bonding

○ Atoms in substances containing only *non-metals* are usually held together by **covalent bonds**.

○ Groups of atoms held together by covalent bonds are called **molecules**.

○ **Diatomic molecules** contain only two atoms – hydrogen, nitrogen, oxygen, the halogens and carbon monoxide exist as diatomic molecules.

○ A covalent bond is the result of two positive nuclei being held together by their common attraction for a shared pair of electrons.

○ *Single*, *double* and *triple* covalent bonds can exist between atoms.

○ A single covalent bond is a pair of electrons shared between two atoms – molecules of hydrogen, the halogens and the hydrogen halides all contain single covalent bonds; for example $H–H$, $Cl–Cl$ and $H–Cl$.

○ A double covalent bond consists of two shared pairs of electrons – one is found in an oxygen molecule, $O=O$.

○ A triple covalent bond is present in a nitrogen molecule, $N\equiv N$.

○ **Molecular formulae** show the number of atoms of the different elements present in a molecule – for example H_2, O_2, N_2, Cl_2, HCl, H_2O and CO_2. These formulae can be written from the molecular models shown in Figure 1.

○ Covalent bonds are strong forces of attraction, but the forces of attraction between discrete molecules are weak. As a result, discrete molecular substances are either gases, liquids or low melting point solids.

○ Covalent substances which exist as discrete molecules do not conduct electricity in any state since the molecules are uncharged.

A closer look at some molecules

○ When two hydrogen atoms share one electron with each other, both obtain the electron arrangement of the noble gas helium (see Figure 2).

○ When a hydrogen atom and a fluorine atom share one electron with each other, hydrogen obtains the electron arrangement of helium and fluorine that of neon (see Figure 3).

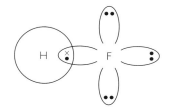

Figure 3 A hydrogen fluoride molecule

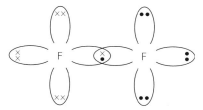

Figure 4 A fluorine molecule

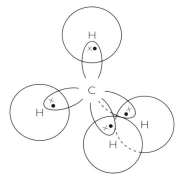

Figure 5 A methane molecule

○ When two fluorine atoms share one electron with each other, each atom obtains the electron arrangement of neon (see Figure 4).

○ Repulsion between the pairs of electrons in the 'lobes' surrounding atoms gives molecules various shapes. A methane molecule has a *tetrahedral* shape (see Figure 5).

○ **Perspective formulae** can be used to show the shapes of molecules. Non-bonded electron pairs, which help to decide the shape of molecules, are not shown in these diagrams. Three common examples are:

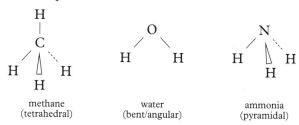

methane (tetrahedral) water (bent/angular) ammonia (pyramidal)

○ **Polar covalent bonds** are formed when the attraction of the atoms for the bonded electrons is different. This results in one end of the bond becoming slightly negative ($\delta-$) and the other end slightly positive ($\delta+$).

○ Examples of molecules containing polar covalent bonds include water, ammonia, hydrogen chloride and hydrogen fluoride. In these molecules, oxygen, nitrogen, chlorine and fluorine atoms attract the bonded electrons more strongly than hydrogen atoms.

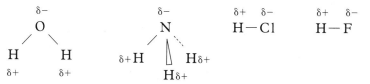

○ Water has unusual properties due to the highly polar nature of its covalent bonds. For example:

– a thin stream of water is attracted strongly to a charged plastic rod

– it has an unusually high boiling point and freezing point when compared with molecules of a similar size.

Covalent network substances

○ Most non-metal elements exist as small discrete molecules – they are **covalent molecular**; for example hydrogen, H_2, and fluorine, F_2.

○ Diamond and graphite are forms of the element carbon which exist as giant lattices of covalently bonded atoms – they are **covalent networks**. A small part of the diamond lattice is shown in Figure 6.

○ Compounds formed between non-metal elements are usually discrete covalent molecular – for example water, H_2O, and carbon dioxide, CO_2.

Figure 6 The diamond structure

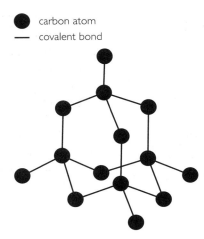

○ Some compounds formed between non-metals are of the covalent network type – for example silicon dioxide, SiO_2, and silicon carbide, SiC. (SiO_2 and SiC are **empirical formulae**, not molecular formulae, since they only show the *ratio* of atoms present in the structure.)

Figure 7 The silicon dioxide structure

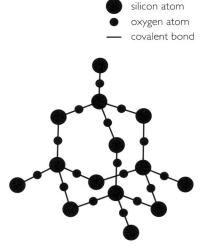

○ Unlike covalent molecular substances, covalent network substances are all solids which have high melting points. This is because, at the melting point, strong covalent bonds must be broken.

○ Like covalent molecular substances, covalent network substances do not conduct electricity in any state, with the exception of carbon in the form of graphite.

Ionic bonding

○ Metal atoms lose electrons to form positively charged **ions** – for example Na^+.

○ Non-metal atoms gain electrons to form negatively charged ions – for example Cl^-.

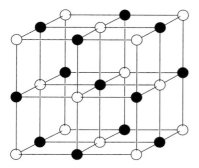

Figure 8 The sodium chloride lattice

◯ An **ionic bond** is the electrostatic force of attraction between oppositely charged ions.

◯ Compounds of metals and non-metals usually contain ionic bonding.

◯ Ionic compounds, such as sodium chloride, exist as giant **ionic lattices** (a small part is shown in Figure 8). Formulae such as Na^+Cl^- give the ratio of ions present in the lattice. (See *Ionic formulae for compounds* on page 28.)

◯ Ionic compounds usually have high melting (and boiling) points because strong ionic bonds have to be broken to melt these compounds.

◯ Ionic compounds do not conduct electricity in the solid state because the ions are not free to move.

◯ Ionic compounds do conduct electricity when molten or dissolved in water because the ions can then move.

◯ An **electrolyte** is a substance that conducts due to the movement of ions.

◯ **Electrolysis** is the flow of ions through solutions or molten compounds accompanied by chemical changes taking place at the **electrodes**.

◯ The literal meaning of 'electrolysis' is 'breaking up by means of electricity'. Electrolysis may lead to the break-up of a compound.

◯ Electrolysis of a molten ionic compound, made up of positive metal ions and negative non-metal ions, results in the metal being produced at the negative electrode and the non-metal at the positive electrode.

◯ Positive metal ions gain electrons at the negative electrode and negative non-metal ions lose electrons at the positive electrode.

◯ Electrolysis of molten lead(II) bromide produces lead and bromine, as shown in Figure 9.

Figure 9 Electrolysis of molten lead(II) bromide

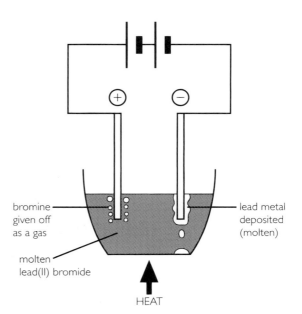

bromine given off as a gas

lead metal deposited (molten)

molten lead(II) bromide

HEAT

At the negative electrode: $Pb^{2+} + 2e^- \rightarrow Pb$
At the positive electrode: $2Br^- \rightarrow Br_2 + 2e^-$

○ A d.c. (direct current) supply must be used for electrolysis experiments; otherwise the signs of the electrodes constantly change and, as a result, the electrode products cannot be identified.

○ Electrolysis of copper(II) chloride solution produces copper and chlorine, as shown in Figure 10.

At the negative electrode: $Cu^{2+} + 2e^- \rightarrow Cu$
At the positive electrode: $2Cl^- \rightarrow Cl_2 + 2e^-$

Figure 10 Electrolysis of coppier(II) chloride solution

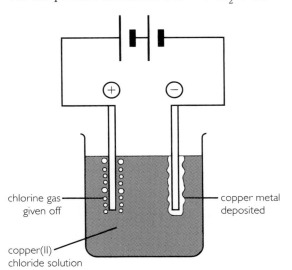

chlorine gas given off

copper metal deposited

copper(II) chloride solution

○ Some ions consist of small groups of atoms joined together – for example chromate ions, CrO_4^{2-}.

○ The colour of an ionic compound can be related to the colours of the ions present – for example sodium chloride is colourless but sodium chromate is yellow, so chromate ions must be yellow.

○ Ion migration experiments can be used to investigate the charges of coloured ions.

Bonding and solubility

○ Ionic compounds are usually soluble in water whereas covalent compounds are usually not.

○ Covalent molecular substances are often soluble in other solvents, such as liquid hydrocarbons like hexane, whereas ionic compounds are usually not.

Metallic bonding

○ Metals consist of a lattice of positively charged ions held together by their electrostatic attraction for delocalised electrons from their outermost energy level, which are free to move throughout the structure.

○ **Metallic bonding** is the electrostatic force of attraction between positive ions in the metallic lattice and the delocalised electrons within the structure.

○ Metallic bonding is usually strong, resulting in most metals having high melting points.

○ The free movement of delocalised electrons within their structures results in metals being good conductors of electricity, as shown in Figure 11.

Figure 11 Conduction of electricity by a metal

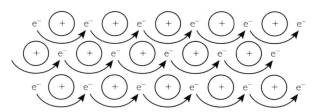

Prescribed Practical Activity

Electrolysis

Safety note

Because of the toxic nature of chlorine, only a tiny amount should be sniffed (by the 'wafting' technique).

- The aim of this experiment is to electrolyse copper(II) chloride solution and to identify the products at the positive and negative electrodes.

- Copper is formed as a brown solid on the negative electrode.

- Chlorine gas, which is poisonous and has a choking smell, is formed at the positive electrode.

- A piece of moist pH paper (or moist blue litmus paper) held above the positive electrode first of all turns red to show that chlorine produces an acidic solution, and is then bleached.

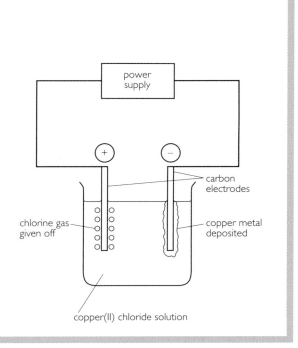

A summary of electrical conduction

○ An electric current is a flow of charged particles.

○ Metals and graphite conduct electricity because some of their electrons are free to move.

○ Ions carry an electric current through electrolytes. These can be either molten ionic compounds or solutions of ionic compounds.

Type of substance	Does conduction take place?		
	Solid state	Liquid state	Solution in water
Ionic	No	Yes	Yes
Covalent	No	No	No
Metallic	Yes	Yes	Insoluble

QUESTIONS

1 Look at these formulae representing different types of substances.

H_2O C H_2
HF   KCl SiO_2

a) Identify the covalent molecular substance which does *not* contain polar covalent bonds.
b) Identify the ionic compound.
c) Identify the *two* covalent molecular substances which contain polar covalent bonds.
d) Identify the substance which can exist as two different covalent network structures.

2 Which of the following compounds contains both a transition metal ion and a halide ion?
A Aluminium bromide
B Cobalt chloride
C Iron oxide
D Sodium fluoride

3 A white solid has a melting point above 1000 °C and dissolves in water. It conducts electricity both in aqueous solution and when molten. It does not conduct in the solid state. The most likely bonding present in the white solid is
A ionic
B polar covalent
C covalent molecular
D covalent network

4 Which of the following exists as diatomic molecules at room temperature?
A Argon
B Sulphur
C Bromine
D Magnesium

5 Refer to the table below which contains information relating to the colour of aqueous solutions.

Solution	Colour
Lithium sulphate	Colourless
Iron(II) sulphate	Pale green
Manganese(II) sulphate	Pale pink
Potassium nitrate	Colourless
Iron(III) chloride	Yellow
Sodium chloride	Colourless

a) What colour would you expect a solution of potassium sulphate to be?
A Pale pink
B Pale green
C Yellow
D Colourless

b) What colour would you expect a solution of manganese(II) chloride to be?
A Pale pink
B Pale green
C Yellow
D Colourless

6 Which of the following correctly describes the particles which carry a current of electricity through copper and copper(II) sulphate solution?

	Copper	Copper(II) sulphate solution
A	Ions	Electrons
B	Electrons	Ions
C	Ions	Ions
D	Electrons	Electrons

7 A sample of lead(II) chloride was melted and electrolysed using a 6 volt d.c. power supply.
   a) Why must a d.c. supply be used in electrolysis experiments?
   b) State what you would expect to see happening at each electrode.
   c) Write ion-electron equations for the reactions taking place at each electrode.

8 Explain the meaning of the following terms:
   a) covalent bond
   b) ionic bond
   c) metallic bond

9 Draw diagrams to show how the outer electrons are shared to form covalent bond(s) in molecules of:
   a) bromine
   b) oxygen
   c) water
   d) phosphine, PH_3

10 Draw perspective formulae (formulae which give symbols and covalent bonds that show the true shapes of the molecules) for:
   a) carbon dioxide
   b) hydrogen sulphide
   c) phosphine, PH_3
   d) tetrachloromethane, CCl_4

11 Both bromine atoms and sulphur atoms attract bonded electrons more strongly than hydrogen atoms. Draw perspective formulae for the following compounds showing the presence of polar covalent bonds by the use of the charge symbols $\delta+$ and $\delta-$:
   a) hydrogen bromide
   b) hydrogen sulphide

12 a) Explain what is meant by the term *polar covalent bond*.
   b) Polar covalent bonds can result in compounds having unusual properties. State three unusual properties which water possesses because of the presence of polar covalent bonds.

13 Carbon dioxide and silicon dioxide have similar chemical formulae (CO_2 and SiO_2), but carbon dioxide is a gas at room temperature whereas silicon dioxide is a hard solid with a high melting point. Explain these differences in properties.

14 In the particle $^{42}_{20}Ca^{2+}$ state the number of protons, neutrons and electrons present, and also give the electron arrangement.

Simple formulae for two-element compounds

Compound formulae from the Periodic Table

○ Simple **formulae** for compounds, both covalent and ionic, indicate the ratio of atoms present.

○ In the case of covalent compounds that exist as molecules, the simple formula is often the molecular formula – for example H_2O.

○ The **valency** of an element is its combining power. The valencies of main group elements in the Periodic Table are as follows:

Group number	1	2	3	4	5	6	7	0
Valency	1	2	3	4	3	2	1	0

○ Simple formulae for two-element compounds can be worked out using the system of valency numbers.

○ By following a step-by-step process, the simple formula for any compound formed between two main group elements can be found.

Example 1

Work out the simple formula for carbon sulphide.

Step 1	symbols	C S
Step 2	valencies	4 2 (carbon is in Group 4, sulphur is in Group 6)
Step 3	cross over the valencies	C_2S_4
Step 4	cancel out any common factor	C_1S_2
Step 5	omit '1' if present	CS_2

The simple formula for carbon sulphide is CS_2.

Example 2

Work out the formula for magnesium bromide.

Step 1	symbols	Mg Br
Step 2	valencies	2 1 (magnesium is in Group 2, bromine is in Group 7)
Step 3	cross over the valencies	Mg_1Br_2
Step 4	cancel out any common factor	Mg_1Br_2
Step 5	omit '1' if present	$MgBr_2$

The simple formula for magnesium bromide is $MgBr_2$.

Compound formulae from prefixes

○ The ratio of atoms present in a compound can be indicated by prefixes.

○ If no prefix is given then it should be assumed that only one atom of that element is present in the formula.

Example 1
Nitrogen monoxide (*mono* = 1)
The simple formula is NO.

Example 2
Sulphur dioxide (*di* = 2)
The simple formula is SO_2.

Example 3
Boron trichloride (*tri* = 3)
The simple formula is BCl_3.

Example 4
Dinitrogen tetraoxide (*tetra* = 4)
The simple formula is N_2O_4.
Note: *penta* = 5 and *hexa* = 6.

Simple formulae for compounds containing group ions

○ *Group ions* contain more than one kind of atom – for example NH_4^+ and CO_3^{2-}. The data booklet contains a table of ions which have more than one kind of atom in them.

○ The valency of ions is the same as the number of charges they have – for example the ammonium ion, NH_4^+, with one positive charge, has a valency of 1; the carbonate ion, CO_3^{2-}, with two negative charges, has a valency of 2.

○ Simple formulae for compounds containing group ions can be worked out using the valency method.

Example 1
Work out the simple formula for ammonium chloride.
Step 1 symbols NH_4 Cl
Step 2 valencies   1   1   (chlorine is in
                                                            Group 7)
Step 3 cross over the valencies $(NH_4)_1Cl_1$
Step 4 cancel out any common factor $(NH_4)_1Cl_1$
Step 5 omit '1' if present NH_4Cl
The simple formula for ammonium chloride is NH_4Cl.

Example 2
Work out the simple formula for sodium carbonate.
Step 1 symbols   Na CO_3
Step 2 valencies   1   2   (sodium is in
                                                            Group 1)
Step 3 cross over the valencies $Na_2(CO_3)_1$
Step 4 cancel out any common factor $Na_2(CO_3)_1$
Step 5 omit '1' if present Na_2CO_3

27

The simple formula for sodium carbonate is Na_2CO_3.

Simple formulae for compounds using roman numerals and brackets

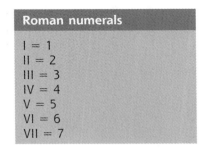

Roman numerals

I = 1
II = 2
III = 3
IV = 4
V = 5
VI = 6
VII = 7

○ Some metals may have more than one valency – for example iron can have a valency of 2 or 3.

○ Chemists use roman numerals to indicate the valency of a metal where this might be in doubt – for example in the compound iron(III) oxide, the valency of iron is 3.

Example 1
Work out the simple formula for iron(III) oxide.
Step 1 symbols   Fe   O
Step 2 valencies   3   2   (oxygen is in Group 6)

Step 3 cross over the valencies Fe_2O_3
Step 4 cancel out any common factor Fe_2O_3
Step 5 omit '1' if present Fe_2O_3
The simple formula for iron(III) oxide is Fe_2O_3.

○ If more than one of the same group ion is present in a formula, then brackets must be used.

Example 2
Work out the simple formula for calcium hydroxide.
Step 1 symbols   Ca   OH
Step 2 valencies   2   1   (calcium is in Group 2 and the hydroxide ion has a single negative charge)

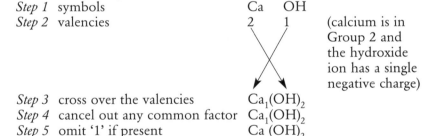

Step 3 cross over the valencies $Ca_1(OH)_2$
Step 4 cancel out any common factor $Ca_1(OH)_2$
Step 5 omit '1' if present $Ca\,(OH)_2$
The simple formula for calcium hydroxide is $Ca\,(OH)_2$.

Ionic formulae for compounds

○ A compound should be assumed to be ionic if it contains a metal or if it is an ammonium compound.

○ When forming an ionic compound, metals in Groups 1, 2 and 3 lose their outer electrons – by doing this they obtain the stable electron arrangement of a noble gas and form positively charged ions.

○ When forming an ionic compound, non-metals in Groups 5, 6 and 7 gain electrons to obtain the stable electron arrangement of a noble gas – by doing this they form negatively charged ions.

○ In Group 4, the non-metals carbon and silicon do not form ions based on single atoms. In the same group, the metals tin and lead show variable valencies.

○ The charges on the ions of main group elements are summarised as follows:

Group number	1	2	3	4	5	6	7	0
Charge on ion	1+	2+	3+	None	3−	2−	1−	None

Example 1

The ionic formula for calcium fluoride, which contains two main group elements (calcium is in Group 2 and fluorine is in Group 7), may be worked out as follows:

Step 1	symbols	Ca	F
Step 2	ions	Ca^{2+}	F^-
Step 3	number of ions to balance charges	$1Ca^{2+}$	$2F^-$ (2+ balances 2−)
Step 4	ionic formula		$Ca^{2+}(F^-)_2$

Note: Whereas in the simple formula, CaF_2, no brackets are required, they *are* required in the ionic formula. Brackets are always needed where more than one ion of a particular type is present in an ionic formula.

○ Where a metal's valency is shown by a roman numeral, this also gives the charge on the metal ion – for example any iron(II) compound will contain the ion Fe^{2+} and any copper(I) compound will contain the ion Cu^+.

○ In the case of ions containing more than one element, the formula is given in the data booklet.

Example 2

The ionic formula for chromium(III) sulphate may be worked out as follows:

Step 1	ions	Cr^{3+}	SO_4^{2-}
Step 2	number of ions to balance charges	$2Cr^{3+}$	$3SO_4^{2-}$
Step 3	ionic formula		$(Cr^{3+})_2(SO_4^{2-})_3$

QUESTIONS

1 The following are two-element compounds where both elements are from a main group in the Periodic Table. It should be noted that the valency of hydrogen is 1. Write simple chemical formulae for:
   a) sodium oxide
   b) calcium oxide
   c) silicon oxide
   d) lithium iodide
   e) barium bromide
   f) boron nitride
   g) lead oxide
   h) tin chloride
   i) aluminium chloride
   j) carbon bromide
   k) magnesium fluoride
   l) magnesium nitride
   m) hydrogen sulphide

n) nitrogen hydride
o) calcium phosphide
p) potassium hydride

2   Use the prefixes in the following names to help you write formulae for:
a) lead monoxide
b) sulphur trioxide
c) chlorine dioxide
d) dinitrogen monoxide
e) phosphorus trichloride
f) carbon tetrachloride
g) tricarbon dioxide
h) sulphur hexafluoride

3   Not every compound has a formula that can be predicted by the valency number method. Which of the compounds referred to in Question 2 have formulae that appear to disobey the rules of the valency number method?

4   In each of the ionic compounds given below there is at least one group ion. You will not need any brackets in your final answers. Write simple chemical formulae for:
a) lithium nitrate
b) calcium carbonate
c) ammonium nitrate
d) calcium chromate
e) barium sulphate
f) sodium hydroxide
g) sodium carbonate
h) caesium hydroxide
i) ammonium iodide
j) barium dichromate
k) lithium phosphate
l) calcium sulphite
m) lithium hydroxide
n) sodium sulphate
o) ammonium nitrite
p) sodium phosphate
q) magnesium sulphate
r) strontium carbonate
s) rubidium nitrate
t) beryllium sulphite
u) potassium nitrate
v) potassium sulphite
w) sodium hydrogencarbonate
x) potassium hydrogensulphite
y) ammonium hydrogensulphate

5   The following compounds contain only two elements. One is a metal with the valency shown by a roman numeral, the other is a non-metal. Write simple chemical formulae for:

a) copper(I) iodide
b) copper(II) oxide
c) iron(III) chloride
d) iron(II) bromide
e) silver(I) oxide
f) gold(III) iodide
g) tin(IV) chloride
h) lead(II) bromide
i) vanadium(V) oxide
j) chromium(III) fluoride
k) cobalt(II) chloride
l) iron(II) sulphide
m) copper(I) chloride
n) silver(I) sulphide
o) zinc(II) sulphide
p) lead(IV) oxide
q) tin(IV) hydride
r) titanium(III) nitride

6   The following series of compounds all contain group ions. You will need brackets in your answers. Write simple chemical formulae for:
a) barium nitrate
b) barium hydroxide
c) calcium phosphate
d) aluminium nitrate
e) magnesium nitrite
f) calcium hydroxide
g) radium phosphate
h) ammonium sulphate
i) strontium nitrate
j) aluminium sulphate
k) aluminium hydroxide
l) strontium phosphate
m) calcium nitrate
n) ammonium carbonate
o) calcium hydrogensulphate
p) magnesium hydrogencarbonate
q) barium hydrogensulphite

7   For the following compounds you require knowledge of the use of roman numerals and/or the use of brackets. Write simple chemical formulae for:
a) silver(I) iodide
b) lead(II) carbonate
c) zinc(II) hydroxide
d) iron(III) sulphate
e) silver(I) chromate
f) cobalt(III) oxide
g) beryllium nitrate
h) iron(III) hydroxide
i) nickel(II) nitrate
j) silver(I) sulphate
k) copper(II) phosphate

l) iron(III) nitrate
m) mercury(I) sulphate
n) ammonium phosphate
o) calcium permanganate
p) magnesium hydrogensulphate
q) copper(II) dichromate

8 The following compounds are made up of two main group elements. Write ionic formulae for:
a) sodium oxide
b) calcium chloride
c) potassium iodide
d) barium sulphide
e) lithium bromide
f) magnesium fluoride
g) beryllium oxide
h) aluminium fluoride
i) potassium phosphide
j) calcium nitride
k) magnesium phosphide

9 In the following compounds at least one of the ions contains more than one element but no brackets are needed. Write ionic formulae for:
a) sodium nitrate
b) potassium hydrogencarbonate
c) calcium carbonate
d) ammonium bromide
e) lithium hydroxide
f) magnesium chromate
g) sodium permanganate
h) calcium dichromate

i) potassium nitrate
j) sodium hydrogensulphate

10 In the following compounds at least one of the ions contains more than one element. Brackets are needed in each case. Write ionic formulae for:
a) sodium carbonate
b) potassium sulphate
c) ammonium carbonate
d) ammonium dichromate
e) lithium phosphate
f) calcium hydrogencarbonate
g) calcium hydroxide
h) aluminium hydroxide
i) barium hydrogensulphite
j) aluminium sulphate
k) ammonium phosphate

11 Write ionic formulae for:
a) copper(I) chloride
b) iron(II) sulphate
c) chromium(III) fluoride
d) silver(I) sulphate
e) iron(III) nitrate
f) copper(II) hydroxide
g) calcium hydrogensulphite
h) iron(III) oxide
i) zinc(II) phosphate
j) ammonium permanganate
k) cobalt(III) fluoride
l) nickel(II) nitrate

Chemical equations

○ In all chemical reactions one or more new substances are formed. The initial substances are called the *reactants* and those formed are called the *products*.

○ **Word equations**, which give only the names of the reactants and products, can be written for reactions – for example

methane + oxygen → carbon dioxide + water

H_2

N_2 | O_2 | F_2

Cl_2

Br_2

I_2

At_2

○ In all word equations '+' between reactants or products means 'and' and '→' means 'react(s) to produce'. Thus, in the above example, methane and oxygen react to produce carbon dioxide and water.

○ Any word equation can be rewritten as a **formula equation** using chemical symbols and formulae for reactants and products.

○ When writing the chemical formula for a diatomic element in a formula equation, the molecular formula is used. The diatomic elements are as shown left.

○ For all other elements, which occur as reactants or products in a formula equation, a single symbol is used – for example C for carbon and Zn for zinc.

31

Example 1
Calcium reacts with water to give calcium hydroxide and hydrogen.

Word equation: calcium + water → calcium hydroxide + water
Formula equation: $Ca + H_2O → Ca(OH)_2 + H_2$

Example 2
When hydrogen burns, it combines with oxygen to produce water.

Word equation: hydrogen + oxygen → water
Formula equation: $H_2 + O_2 → H_2O$

○ The formulae equations shown above are unbalanced. Balancing equations is dealt with in the next section.

QUESTIONS

1 When a substance burns it combines with oxygen to produce compounds called oxides (in most cases). Use this information to write formulae equations for:
   a) sodium burning to produce sodium oxide
   b) calcium burning to produce calcium oxide
   c) phosphorus burning to produce diphosphorus trioxide
   d) sulphur burning to produce sulphur dioxide
   e) propane (C_3H_8) burning to give carbon dioxide and water

2 Write formulae equations for each of the following word equations:
   a) hydrogen + fluorine → hydrogen fluoride
   b) mercury + iodine → mercury(II) iodide
   c) iron + copper(II) sulphate → copper + iron(II) sulphate
   d) magnesium + zinc(II) nitrate → zinc + magnesium nitrate
   e) barium + water → barium hydroxide + hydrogen

3 Write formulae equations for each of the following sentence descriptions.
   a) Calcium hydroxide and carbon dioxide react to give calcium carbonate and water
   b) Magnesium carbonate reacts with water and carbon dioxide to give magnesium hydrogencarbonate
   c) Barium oxide and water react to give barium hydroxide
   d) Lead(II) nitrate and potassium iodide react to give lead(II) iodide and potassium nitrate
   e) Sodium hydroxide and sulphur dioxide react to give sodium sulphite and water

Balancing chemical equations

○ A **balanced chemical equation** is needed in order to carry out calculations – for example to predict the mass of product in a reaction.

○ A chemical equation is said to be *balanced* when there are equal numbers of each type of atom on both sides of the equation. This is logical since atoms are not made or destroyed during chemical reactions – all the atoms which were present in the reactants are still present in the products.

○ Equations should only be balanced by putting numbers, as required, in front of formulae – formulae must not be altered in any way when balancing equations.

○ It is usually advisable to start with the first element on the left-hand side and to continue methodically through the equation.

Carry out a final check to make sure you have not unbalanced any elements.

Example 1

In the complete combustion of methane, methane and oxygen react to produce carbon dioxide and water. The word equation for the reaction is:

$$\text{methane} + \text{oxygen} \rightarrow \text{carbon dioxide} + \text{water}$$

Writing formulae in place of the names of the reactants and products gives the *unbalanced* formulae equation:

$$CH_4 + O_2 \rightarrow CO_2 + H_2O$$

Although there is one carbon atom on both sides of the equation, there are four hydrogen atoms on the left, but only two on the right. There are also two oxygen atoms on the left, but three on the right. This can be seen more clearly if full structural formulae are used:

$$
\begin{array}{c}
\quad\quad H \\
\quad\quad | \\
H-C-H \;+\; O{=}O \;\rightarrow\; O{=}C{=}O \;+\; H-O \\
\quad\quad | \quad\quad\quad\quad\quad\quad\quad\quad\quad\quad\quad\quad\;\; \backslash H \\
\quad\quad H
\end{array}
$$

Left:		Right:	
H atoms 4		H atoms 2	
C atoms 1		C atoms 1	
O atoms 2		O atoms 3	

The four hydrogen atoms in the methane molecule will produce two water molecules, each containing two hydrogen atoms. Two oxygen molecules will be required to provide the four oxygen atoms that are present in the products:

$$
\begin{array}{c}
\quad\quad H \\
\quad\quad | \quad\quad\quad\quad O{=}O \quad\quad\quad\quad\quad\quad\quad\quad\quad H-O \\
H-C-H \;+ \quad\quad\quad\quad\;\rightarrow\; O{=}C{=}O \;+ \quad\quad\quad\; \backslash H \\
\quad\quad | \quad\quad\quad\quad O{=}O \quad\quad\quad\quad\quad\quad\quad\quad\quad H-O \\
\quad\quad H \quad\quad\quad\quad\quad\quad\quad\quad\quad\quad\quad\quad\quad\quad\quad\quad\quad\quad \backslash H
\end{array}
$$

Left:		Right:	
H atoms 4		H atoms 4	
C atoms 1		C atoms 1	
O atoms 4		O atoms 4	

The balanced equation can now be written as:

$$CH_4 + 2O_2 \rightarrow CO_2 + 2H_2O$$

Example 2

Balance the equation:

$$H_2 + O_2 \rightarrow H_2O$$

The hydrogen atoms are balanced in this case (two atoms on each side), so proceed to oxygen. There are two oxygen atoms on the left, but only one on the right, so we put a '2' in front of H_2O. This gives: **33**

$$H_2 + O_2 \rightarrow 2H_2O$$

However, checking through again, we find that there are two hydrogen atoms on the left, but four on the right. In order to have four hydrogen atoms on the left, we put a '2' in front of H_2. This gives:

$$2H_2 + O_2 \rightarrow 2H_2O$$

A final check shows that the equation is now balanced:

Left: H atoms 4   Right: H atoms 4
           O atoms 2   O atoms 2

Sometimes the use of a fraction, such as ½, can help to balance an equation.

Example 3

Balance the equation:

$$C_2H_6 + O_2 \rightarrow CO_2 + H_2O$$

Balancing the carbon atoms gives:

$$C_2H_6 + O_2 \rightarrow 2CO_2 + H_2O$$

Balancing the hydrogen atoms gives:

$$C_2H_6 + O_2 \rightarrow 2CO_2 + 3H_2O$$

There are now seven oxygen atoms on the right so the number to put in front of O_2 must give seven when multiplied by the '2' in O_2. This number is 3½ which gives:

$$C_2H_6 + 3\tfrac{1}{2}O_2 \rightarrow 2CO_2 + 3H_2O$$

Left: C atoms 2   Right: C atoms 2
           H atoms 6   H atoms 6
           O atoms 7   O atoms 7

The equation is now balanced and is quite acceptable.

Multiplying through by 2 gives the simplest possible whole numbers, which some prefer. The equation then becomes:

$$2C_2H_6 + 7O_2 \rightarrow 4CO_2 + 6H_2O$$

Left: C atoms 4   Right: C atoms 4
           H atoms 12   H atoms 12
           O atoms 14   O atoms 14

❍ There are rare occasions when the methodical approach described in Examples 2 and 3 does not lead to a balanced equation. In such cases a 'trial and error' method may be adopted, or it may be possible to solve the problem by a closer examination of the particular equation.

Example 4

Consider the balancing of the equation:

$$Fe_2O_3 + CO \rightarrow Fe + CO_2$$

Balancing the iron atoms gives:

$$Fe_2O_3 + CO \rightarrow 2Fe + CO_2$$

Trying to balance the carbon and oxygen atoms appears to be impossible by the usual method. However, it should be noticed that each carbon dioxide molecule removes one oxygen atom from the iron(III) oxide. There are three oxygen atoms to be removed and this, therefore, requires three carbon monoxide molecules. In turn, this leads to the production of three carbon dioxide molecules. The balanced equation is therefore:

$$Fe_2O_3 + 3CO \rightarrow 2Fe + 3CO_2$$

Left: Fe atoms 2   Right: Fe atoms 2
        O atoms 6   O atoms 6
        C atoms 3   C atoms 3

QUESTIONS

Balance the following equations. In each case give the simplest possible *whole numbers* in your final answer. Some equations are already balanced.

1 $Na + F_2 \rightarrow NaF$

2 $Mg + Cl_2 \rightarrow MgCl_2$

3 $N_2 + H_2 \rightarrow NH_3$

4 $K + O_2 \rightarrow K_2O$

5 $H_2 + Br_2 \rightarrow HBr$

6 $Zn + H_2SO_4 \rightarrow ZnSO_4 + H_2$

7 $Ca + HCl \rightarrow CaCl_2 + H_2$

8 $Li_2O + H_2O \rightarrow LiOH$

9 $SnO_2 + H_2 \rightarrow Sn + H_2O$

10 $KOH + H_2SO_3 \rightarrow K_2SO_3 + H_2O$

11 $Ca(OH)_2 + HCl \rightarrow CaCl_2 + H_2O$

12 $Ba(OH)_2 + H_2SO_4 \rightarrow BaSO_4 + H_2O$

13 $Cu(NO_3)_2 \rightarrow CuO + NO_2 + O_2$

14 $MgCO_3 + HCl \rightarrow MgCl_2 + H_2O + CO_2$

15 $C_3H_6 + O_2 \rightarrow CO_2 + H_2O$

16 $C_6H_{14} + O_2 \rightarrow CO_2 + H_2O$

17 $C_5H_{10} + O_2 \rightarrow CO_2 + H_2O$

18 $Fe_3O_4 + CO \rightarrow Fe + CO_2$

19 $PCl_3 + H_2O \rightarrow H_3PO_3 + HCl$

20 $NH_4NO_3 \rightarrow N_2O + H_2O$

(1.6) The mole

Formula mass and the mole

○ The **formula mass** of a substance is obtained by adding the relative atomic masses of the elements in the formula. The number of times a symbol occurs must be taken into account.

Example 1
Calculate the formula mass of calcium chloride.
Formula $CaCl_2$ (RAMs: Ca = 40; Cl = 35.5)
Formula mass $40 + (35.5 \times 2) = 111$

Example 2
Calculate the formula mass of ammonium sulphate.
Formula $(NH_4)_2SO_4$ (RAMs: N = 14; H = 1; S = 32; O = 16)
Formula mass $[14 + (1 \times 4)] \times 2 + 32 + (16 \times 4) = 132$

○ A **mole** of a substance is the formula mass in grams – for example the mass of one mole of calcium chloride is 111 g and the mass of one mole of ammonium sulphate is 132 g.

○ For any element or compound, the mass present can be calculated from a knowledge of the number of moles and the formula mass:

mass of substance = number of moles × mass of one mole
= number of moles × formula mass in grams

Mass of substance in grams = number of moles × formula mass

Example 3
Calculate the mass of 2.5 moles of calcium carbonate, $CaCO_3$.
Mass of $CaCO_3$ in grams   = number of moles × formula mass
= $2.5 \times [40 + 12 + (16 \times 3)] = 250$ g

○ Rearranging the relationship above gives:

$$\text{number of moles of substance} = \frac{\text{mass of substance in grams}}{\text{formula mass}}$$

From this we can calculate the number of moles of substance present, if we know the mass of substance in grams and the formula mass.

Example 4
Calculate the number of moles of water in 100 g of water.

$$\text{Number of moles of } H_2O = \frac{\text{mass of } H_2O \text{ in grams}}{\text{formula mass}}$$

$$= \frac{100}{[(1 \times 2) + 16]}$$

$$= \frac{100}{18} = 5.56$$

○ The relationship between the *mass of substance in grams* (*m*), the *number of moles of substance* (*n*) and the *formula mass* (*fm*) can be summed up in the triangle of knowledge:

QUESTIONS

1 Calculate the formula mass of each of the following substances:
  a) oxygen, O_2
  b) sulphur dioxide, SO_2
  c) ammonia, NH_3
  d) butane, C_4H_{10}
  e) sodium carbonate, Na_2CO_3
  f) calcium hydroxide, $Ca(OH)_2$
  g) ammonium phosphate, $(NH_4)_3PO_4$
  h) magnesium nitrate, $Mg(NO_3)_2$
  i) aluminium sulphate, $Al_2(SO_4)_3$

2 Calculate the mass of each of the following:
  a) 1 mole of hydrogen, H_2
  b) 1 mole of carbon dioxide, CO_2
  c) 3 moles of ammonia, NH_3
  d) 2.5 moles of sodium hydroxide, NaOH
  e) 5 moles of lithium oxide, Li_2O
  f) 0.25 moles of potassium sulphate, K_2SO_4
  g) 0.15 moles of iron(III) nitrate, $Fe(NO_3)_3$
  h) 1.5 moles of ammonium dichromate, $(NH_4)_2Cr_2O_7$ (RAM of Cr = 51)

3 Calculate the number of moles of each of the following:
  a) 98 g of nitrogen, N_2

b) 280 g of potassium hydroxide, KOH
c) 92.5 g of lithium carbonate, Li_2CO_3
d) 500 g of ammonium nitrate, NH_4NO_3
e) 3.68 g of calcium nitrate $Ca(NO_3)_2$
f) 4 kg of magnesium hydrogensulphate, $Mg(HSO_4)_2$ (1 kg = 1000 g)
g) 2.5 kg of calcium phosphate, $Ca_3(PO_4)_2$
h) 10 kg of copper (II) hydroxide, $Cu(OH)_2$

4 An indigestion remedy contains 0.5 g of calcium carbonate, $CaCO_3$, in each tablet. It is recommended that no more than 16 tablets are taken in one day.
How many moles of calcium carbonate are present in 16 tablets?

5 A gas cylinder used in caravans contains 10 kg of butane, C_4H_{10}.
How many moles of butane are in the cylinder?

6 A company can make 100 tonnes of the fertiliser ammonium phosphate in one day. How many moles of ammonium phosphate does this represent? (1 tonne = 1000 kg)

Calculations based on balanced equations

○ A balanced chemical equation tells us the number of moles of each reactant and product in a given reaction.

○ The mass of a reactant or product can be calculated using a balanced equation.

○ A calculation based on a balanced equation can be broken down into five stages:

Step 1 write a balanced equation for the reaction

Step 2 identify the number of moles of the substances concerned

Step 3 replace moles by formula masses in grams

Step 4   change grams to any other unit of mass if necessary
Step 5   use simple proportion to complete the calculation.

Example 1 (no change of units required)

Calculate the mass of magnesium oxide produced when 6 g of magnesium burns in air or oxygen.

$$2Mg \quad + \quad O_2 \quad \rightarrow \quad 2MgO$$

2 mol $\longrightarrow$ 2 mol

48 g $\longrightarrow$ 80 g

6 g $\longrightarrow$ $\dfrac{80 \times 6}{48} = 10$ g

Example 2 (change of units required)

What mass of water is produced when 10 tonnes of methane burns completely in air?

$$CH_4 \quad + \quad 2O_2 \quad \rightarrow \quad CO_2 \quad + \quad 2H_2O$$

1 mol $\longrightarrow$ 2 mol

16 g $\longrightarrow$ 36 g

16 tonnes $\longrightarrow$ 36 tonnes

10 tonnes $\longrightarrow$ $\dfrac{36 \times 6}{48} = 22.5$ tonnes

QUESTIONS

1 The charcoal used in some barbecues is mainly carbon. Calculate the mass of carbon dioxide produced when 480 g of carbon burns.

$$C + O_2 \rightarrow CO_2$$

2 The halogens can combine directly with metals. What mass of sodium chloride is formed when 2 g of sodium reacts with chlorine?

$$2Na + Cl_2 \rightarrow 2NaCl$$

3 Lime (calcium oxide) is formed when limestone (calcium carbonate) is heated. What mass of lime is produced by the complete decomposition of 5 tonnes of limestone?

$$CaCO_3 \rightarrow CaO + CO_2$$

4 The reaction between calcium carbonate and dilute hydrochloric acid is often used in school laboratories for the production of carbon dioxide. What mass of carbon dioxide would be produced by the complete reaction of 20 g of calcium carbonate?

$$CaCO_3 + 2HCl \rightarrow CaCl_2 + H_2O + CO_2$$

5 A single serving of a breakfast cereal contains 52 mg of iron which, in the stomach, reacts with hydrochloric acid to produce hydrogen. What mass of hydrogen would be produced in this reaction?

$$Fe + 2HCl \rightarrow FeCl_2 + H_2$$

6 Potassium nitrate, which is found in gunpowder, can be made by a reaction between potassium hydroxide and nitric acid. What mass of potassium nitrate could be formed by the complete reaction of 15 kg of potassium hydroxide?

$$KOH + HNO_3 \rightarrow KNO_3 + H_2O$$

7 A company makes ammonium sulphate fertiliser by a reaction between ammonia and sulphuric acid. What mass of ammonia would be required to make 500 kg of ammonium sulphate?

$$2NH_3 + H_2SO_4 \rightarrow (NH_4)_2SO_4$$

8 Some iron ore is almost pure iron(III) oxide. What mass of this compound would be required in order to produce 10 tonnes of iron in a blast furnace?

$$Fe_2O_3 + 3CO \rightarrow 2Fe + 3CO_2$$

9 Astronauts travelling to the moon drank water produced from hydrogen in fuel cells. What mass of hydrogen would be required to produce 1000 cm^3 of water? (1 cm^3 of water has a mass of 1 g.)

$$2H_2 + O_2 \rightarrow 2H_2O$$

(1.7) Glossary

1.1

alkali metals   Reactive metals in Group 1 of the Periodic Table (Li, Na, K, Rb, Cs, Fr).

chemical reaction   A chemical process in which one or more new substances are formed – usually accompanied by an energy change and a change in appearance.

chromatography   A method for separating small quantities of soluble compounds. (Refers to paper chromatography.)

compound   A substance in which two or more elements are joined together chemically.

concentrated solution   One with a lot of solute compared with the amount of solvent.

dilute solution   One with little solute compared with the amount of solvent.

distillation   A process of separation or purification dependent on differences in boiling point. The changes of state involved are liquid to gas to liquid.

element   A substance which cannot be broken down into simpler substances by chemical means. All of its atoms have the same atomic number.

endothermic reaction   One in which heat energy is taken in from the surroundings causing a decrease in temperature.

evaporation   The process of turning a liquid into a vapour, usually by heating.

exothermic reaction   One in which heat energy is given out to the surroundings causing an increase in temperature.

filtrate   The liquid which passes through filter paper during filtration.

filtration   The separation of an insoluble solid from a liquid by passage through filter paper.

group   A column of elements in the Periodic Table.

halogens   Reactive non-metals in Group 7 of the Periodic Table (F, Cl, Br, I, At).

metals   Shiny, malleable elements found on the left-hand side of the Periodic Table. They all conduct electricity.

mixture   Two or more substances mixed together but not chemically joined – for example air and crude oil.

noble gases   Very unreactive gases in Group 0 of the Periodic Table (He, Ne, Ar, Xe, Kr, Rn).

non-metals   Non-conducting elements (except carbon as graphite) found on the right-hand side of the Periodic Table.

period   A horizontal row in the Periodic Table.

Periodic Table   An arrangement of the elements in order of increasing atomic number, with chemically similar elements occurring in the same vertical columns (groups).

residue   The solid left behind in the filter paper after filtration.

saturated solution   Solution in which no more solute will dissolve at a given temperature.

solute   A substance that dissolves in a liquid.

solution   A liquid mixture formed when a solute dissolves in a solvent.

solvent   A liquid in which a substance dissolves.

state symbols   Symbols used to indicate the state of a substance or species:

(s) = solid; (l) = liquid; (g) = gas; (aq) = aqueous (dissolved in water).

1.2

adsorption   The attachment of molecules of a liquid or gas to the surface of another substance – usually a solid.

catalyst   A substance which speeds up a chemical reaction but is not used up and can be recovered unchanged at the end of the reaction.

catalyst poison   Something which destroys the activity of a catalyst.

catalytic converter   Part of the exhaust system of a car where platinum and other catalysts change the pollutants carbon monoxide, oxides of nitrogen and unburned hydrocarbons into less harmful carbon dioxide, nitrogen and water vapour.

enzyme   A biological catalyst.

heterogeneous catalyst   One which is in a different state from the reactants.

homogeneous catalyst   One which is in the same state as the reactants.

rate of reaction   A measure of how quickly a chemical reaction takes place – how rapidly reactants are used up and products are produced.

variable   Something that can be changed in a chemical reaction – for example temperature, particle size and concentration.

1.3

atom   The smallest part of an element that can exist. It has a nucleus of protons and (usually) neutrons surrounded by moving electrons.

atomic number   The number of protons in the nucleus of an atom.

electron   A particle which moves around the nucleus of an atom. It has a single negative charge but its mass is negligible compared with that of a proton or neutron.

energy levels (shells)   Where electrons are found, outside the nucleus in an atom.

isotopes   Atoms of the same element which have different numbers of neutrons. They have the same atomic number but different mass numbers.

mass number   The total number of protons and neutrons in the nucleus of an atom.

neutron   A particle found in the nucleus of an atom. It has the same mass as a proton but no charge.

nucleus   The extremely small centre of an atom where the protons and neutrons are found.

proton   A particle found in the nucleus of all atoms. It has a single positive charge and the same mass as a neutron.

relative atomic mass   The average mass of one atom of an element on a scale where one atom of ^{12}C has a mass of 12 units exactly. It is the average of the mass numbers of an element's isotopes taking into account the proportions of each.

1.4

covalent bond   The result of two nuclei being held together by their common attraction for a shared pair of electrons (single bond); or two pairs of electrons (double bond) or three pairs of electrons (triple bond).

covalent molecular   Existing as small discrete molecules, for example hydrogen and fluorine.

covalent network   A giant lattice of atoms held together by covalent bonds – for example diamond, graphite and silicon dioxide.

diatomic element   An element whose molecules contain only two atoms – for example H_2, N_2, O_2, F_2, Cl_2, Br_2, I_2, At_2.

diatomic molecule   A molecule which contains only two atoms – for example H_2 and HCl.

electrode   A conductor which is used to pass electricity into and out of solutions and melts – for example carbon as graphite.

electrolysis   The flow of ions through electrolytes (ionic solutions or molten ionic compounds) accompanied by chemical changes at the electrodes which can result in the decomposition of the electrolyte.

electrolyte   A compound which conducts due to the movement of ions either when dissolved in water or melted.

empirical formula   Shows the simplest ratio of atoms in a compound – for example SiO_2 is the empirical formula for silicon dioxide.

ion   Atom or group of atoms which possess a positive or negative charge due to loss or gain of electrons – for example Na^+ and Cl^-.

ionic bond   The electrostatic force of attraction between oppositely charged ions.

ionic lattice   A giant arrangement of ions held together by electrostatic attraction (ionic bonds) – for example sodium chloride.

metallic bond   The electrostatic force of attraction between positive ions in a metallic lattice and the delocalised electrons within the structure.

molecular formula   A formula which shows the number of atoms of different elements present in one molecule of a substance.

molecule   A group of atoms held together by covalent bonds.

perspective formula   A formula which shows the true shape of a molecule.

polar covalent bond   A covalent bond in which the electron pair is not shared equally making one atom slightly positive ($\delta+$) and the other slightly negative ($\delta-$).

1.5

balanced chemical equation   One with the same number of atoms of each element on both sides of the equation.

formula   A simple chemical formula for a molecular substance – usually indicates the number of atoms of each element in a molecule – for example CO_2. In other cases the ratio of atoms or ions present is shown – for example NaCl.

formulae equation   An equation which gives chemical symbols and formulae for reactants and products.

ionic formula   One which shows any ions which may be present among the reactants and products.

valency   A number which indicates the 'combining power' of atoms or ions.

word equation   An equation which gives only the names of reactants and products.

1.6

formula mass   The sum of the relative atomic masses of all the atoms present in a formula.

mole   A formula mass expressed in grams.

CARBON COMPOUNDS

(2.1) Fuels

Combustion

○ A **fuel** is a substance which is burned to produce energy.

○ An **exothermic** reaction is one in which heat energy is released.

○ When a substance burns it reacts with oxygen.

○ Oxygen makes up about one-fifth of the air – the rest is mainly nitrogen.

○ The test for oxygen is that it relights a glowing splint.

○ **Combustion** is another word for burning.

Figure 1 Combustion of a hydrocarbon fuel to give carbon dioxide and water

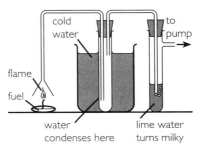

○ Crude oil and natural gas are mixtures consisting mainly of carbon compounds.

○ The compounds found in crude oil and natural gas are mainly **hydrocarbons**.

○ A hydrocarbon is a compound which contains carbon and hydrogen only.

○ Hydrocarbons burn in a plentiful supply of air to produce carbon dioxide and water – for example methane + oxygen → carbon dioxide + water

$$CH_4 + 2O_2 \rightarrow CO_2 + 2H_2O$$

○ The test for carbon dioxide is that it turns **lime water** milky.

○ The test for water is that it freezes at $0\,°C$ and boils at $100\,°C$.

○ The production of carbon dioxide and water on combustion indicates the presence of carbon and hydrogen in the original fuel. (Only the oxygen in these compounds came from the air.) **43**

- ◯ Carbon dioxide is a 'greenhouse' gas that causes global warming.
- ◯ Incomplete combustion of hydrocarbons, or any fuel which contains carbon, can produce carbon and carbon monoxide.
- ◯ Carbon monoxide is a poisonous gas.
- ◯ Soot particles (mainly carbon) produced by the incomplete combustion of the hydrocarbons in diesel are harmful.
- ◯ In petrol engines, and to a lesser extent in diesel engines, nitrogen and oxygen from the air react to form poisonous oxides of nitrogen, including nitrogen dioxide. In petrol engines, this is caused by the high temperature reached near the spark plug.
- ◯ Some fuels contain sulphur, which produces poisonous sulphur dioxide when burned. Most sulphur is removed from crude oil and natural gas during refining and processing.
- ◯ Air pollution from the burning of hydrocarbons can be reduced by the use of **catalytic converters** which change pollutant gases into harmless gases. Oxides of nitrogen, carbon monoxide along with unburned hydrocarbons are changed into nitrogen, carbon dioxide and water vapour.

Fractional distillation of crude oil

- ◯ **Fractional distillation** is used to separate crude oil into **fractions**.
- ◯ A fraction is a group of hydrocarbons with boiling points within a given range.

Fraction	Boiling range/°C	Carbon atoms per molecule	End uses
Gas	<20	1–4	Fuel gases
Gasoline Naphtha	20–65 65–180	5–6 6–11	Petrol and petrochemicals
Kerosene	180–250	9–15	Heating/jet fuel
Gas oils	250–350	15–25	Diesel fuel
Residue	>350	>25	Bitumen, wax etc.

- ◯ The fractions vary in **viscosity** (how 'thick' it is), **flammability** and ease of **vaporisation.**

Fraction	Gas Gasoline	Naphtha	Kerosene Gas/oil Residue
Molecular size	———————	increasing	———————→
Boiling range	———————	increasing	———————→
Viscosity	———————	increasing	———————→
Flammability	←———————	increasing	———————
Vaporisability	←———————	increasing	———————

○ Boiling range and viscosity increase with increasing molecular size because the forces of attraction between the molecules increase.

○ Ease of vaporisation ('vaporisability') and flammability are also linked to molecular size. The smaller molecules have weaker forces of attraction between them and, as a result, they vaporise more readily than larger molecules.

○ The uses of the fractions are related to their ease of vaporisation, viscosity, flammability and boiling range.

QUESTIONS

1 These substances are products from an oil refinery.

fuel gas	diesel	petrol
bitumen	kerosene	lubricating oil

a) Which product could consist of a mixture of hydrocarbons having **(i)** more than 30 carbon atoms in each molecule, **(ii)** three or four carbon atoms in each molecule?

b) Which product is more flammable than diesel but less flammable than petrol?

c) Which product is a liquid which is less viscous than kerosene?

2 Which of the following hydrocarbons is a constituent of petrol?
A C_2H_6
B C_7H_{16}
C $C_{13}H_{28}$
D $C_{17}H_{36}$

3 The only products of the complete combustion of a fuel were shown to be carbon dioxide and water. The fuel therefore contained:
A   carbon, hydrogen and oxygen
B   hydrogen because carbon and oxygen are present in the air
C   carbon because hydrogen and oxygen are in the air
D   carbon and hydrogen because oxygen is in the air

4 Which of the following gases is *not* considered to be an active poison?

A   carbon dioxide
B   sulphur dioxide
C   carbon monoxide
D   nitrogen dioxide

5 Fuels are said to react exothermically with oxygen.
a) State the meaning of the terms *fuel* and *exothermic reaction*.
b) Describe the test for oxygen.

6 Crude oil and natural gas are mainly made up of hydrocarbons. What is meant by the word *hydrocarbon*?

7 Diesel engines and petrol engines both burn hydrocarbon fuels, but the products of combustion tend to be different.
a) Which harmful solid particles can be produced by the incomplete combustion of diesel?
b) Suggest a reason why diesel engines produce less nitrogen dioxide than petrol engines.

8 Copy and complete:
The continuous process of _____ _____ is used to separate crude oil into fractions. A _____ is a group of hydrocarbons with boiling _____ within a given range. Viscosity _____ with increase in molecular size because the forces of _____ between the molecules also _____.

Hydrocarbons

Alkanes

○ The **alkanes** are a subset of the set of hydrocarbons. They are the main constituents of crude oil and natural gas. They can be identified from the '–ane' ending of their names. The eight simplest are:

Name	Molecular formula and state at room temperature	Shortened structural formula
Methane	$CH_4(g)$	CH_4
Ethane	$C_2H_6(g)$	CH_3CH_3
Propane	$C_3H_8(g)$	$CH_3CH_2CH_3$
Butane	$C_4H_{10}(g)$	$CH_3CH_2CH_2CH_3$
Pentane	$C_5H_{12}(l)$	$CH_3CH_2CH_2CH_2CH_3$
Hexane	$C_6H_{14}(l)$	$CH_3CH_2CH_2CH_2CH_2CH_3$
Heptane	$C_7H_{16}(l)$	$CH_3CH_2CH_2CH_2CH_2CH_2CH_3$
Octane	$C_8H_{18}(l)$	$CH_3CH_2CH_2CH_2CH_2CH_2CH_2CH_3$

○ The **general formula** for the alkanes is C_nH_{2n+2}.

○ Full **structural formulae** are often used for the alkanes – for example

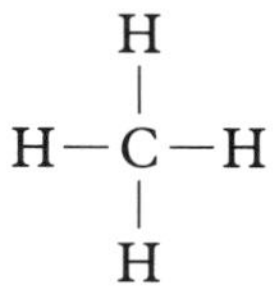

$$
\begin{array}{ccc}
\quad\;\; H & \quad H \;\; H & \quad\; H \;\;\; H \;\;\; H \\
\quad\;\; | & \quad\;\; | \;\;\; | & \quad\;\; | \;\;\; | \;\;\; | \\
H-C-H & H-C-C-H & H-C-C-C-H \\
\quad\;\; | & \quad\;\; | \;\;\; | & \quad\;\; | \;\;\; | \;\;\; | \\
\quad\;\; H & \quad H \;\; H & \quad\; H \;\;\; H \;\;\; H \\
\text{methane} & \text{ethane} & \text{propane}
\end{array}
$$

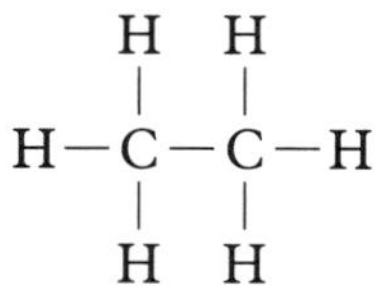

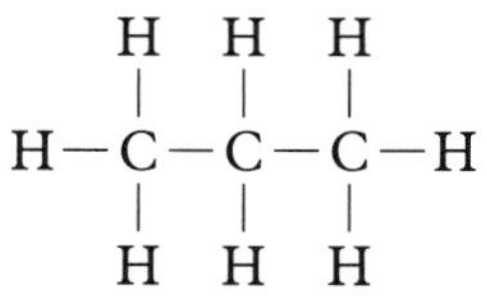

Branched chain alkanes

○ The alkanes, methane, ethane, propane, etc., are described as *straight chain*, with each carbon atom being joined directly to no more than two others.

○ In *branched chain* alkanes, some carbon atoms are joined directly to three or four others.

○ The rules for naming branched chain alkanes are:

– look for the longest carbon chain to give the name of the parent alkane

- number the carbon atoms in the longest chain from the end that gives the lowest numbers to the carbon atoms with branches attached
- identify the groups making up the branches: CH_3 is methyl, CH_2CH_3 is ethyl, $CH_2CH_2CH_3$ is propyl, ...
- use the prefixes di-, tri-, ... to indicate how many of a particular side group are present
- indicate the position of each side group with a separate number placed in front of its name
- if more than one type of side group is present, the names are given in alphabetical order.

For example, 2,2,4-trimethylpentane

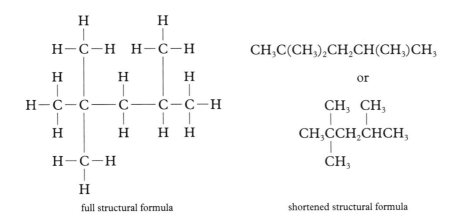

$CH_3C(CH_3)_2CH_2CH(CH_3)CH_3$

or

CH_3 CH_3
$CH_3CCH_2CHCH_3$
CH_3

full structural formula   shortened structural formula

For example, 2,4-dimethyl-4-ethylhexane

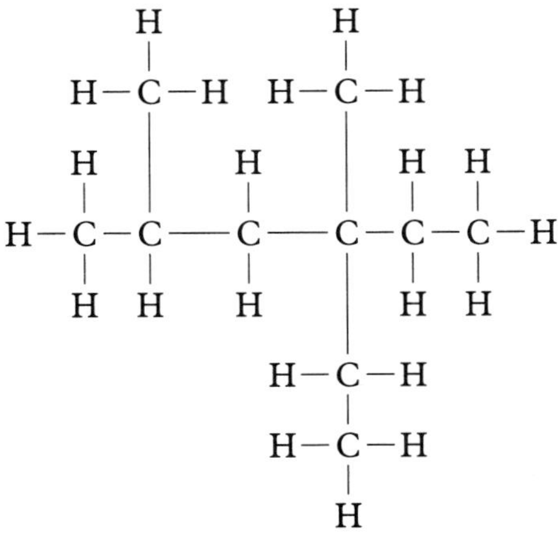

$CH_3CH(CH_3)CH_2C(CH_3)(C_2H_5)CH_2CH_3$

or

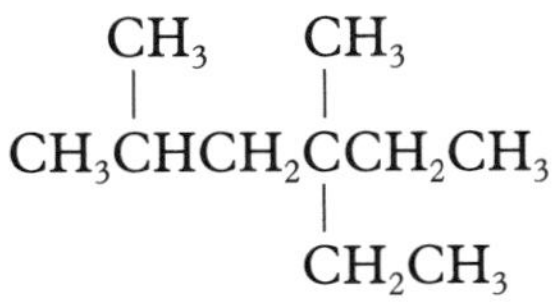

CH_3 CH_3
$CH_3CHCH_2CCH_2CH_3$
CH_2CH_3

full structural formula   shortened structural formula

○ In a branched chain alkane the longest carbon chain is not always shown in a *linear* manner. For example, the name of the molecule shown below is

3-methylhexane, not 2-ethylpentane.

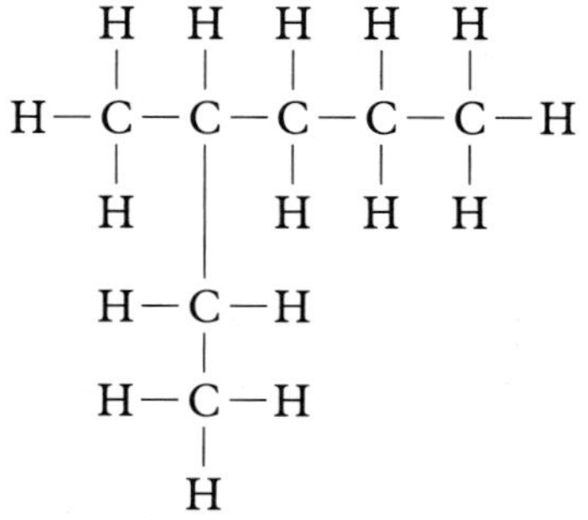

$$CH_3CH(C_2H_5)CH_2CH_2CH_3$$

or

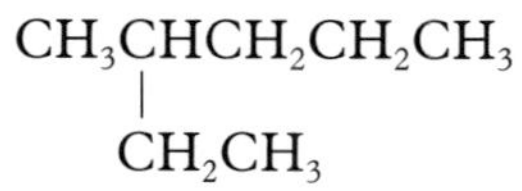

$$CH_3CHCH_2CH_2CH_3$$
$$CH_2CH_3$$

full structural formula   shortened structural formula

Alkenes

○ The **alkenes** is another subset of the set of hydrocarbons. They are similar to the alkanes, but contain a double $C = C$ bond, indicated by the '–ene' name ending. Examples include:

Name	Molecular formula and state at room temperature	Shortened structural formula
Ethene	$C_2H_4(g)$	$CH_2 = CH_2$
Propene	$C_3H_6(g)$	$CH_2 = CHCH_3$
But-1-ene	$C_4H_8(g)$	$CH_2 = CHCH_2CH_3$
But-2-ene	$C_4H_8(g)$	$CH_3CH = CHCH_3$

○ The pentenes are the first alkenes that are liquids at room temperature.
○ The general formula for the alkenes is C_nH_{2n}.
○ Full structural formulae are often used for the alkenes – for example

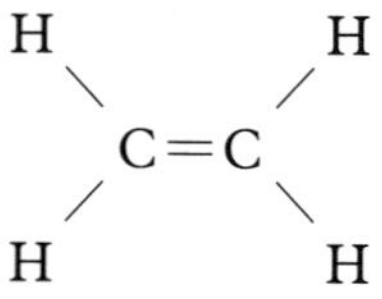

ethene 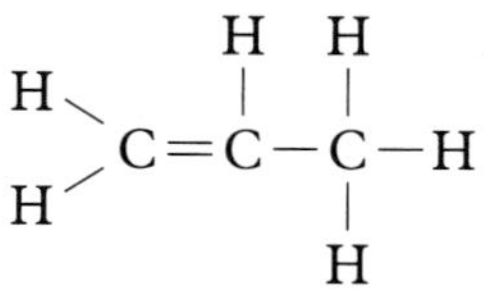   propene 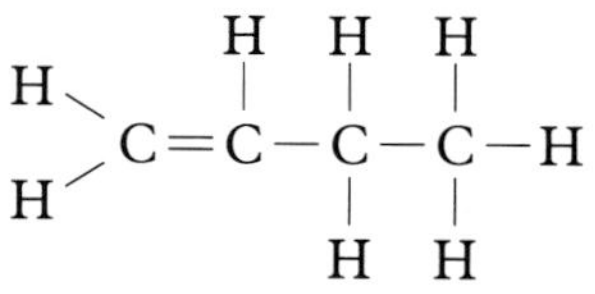   but-1-ene

○ The rules for naming alkenes are:
  – look for the longest carbon chain containing the $C = C$ bond to give the name of the parent alkane
  – remove the -ane ending of the parent alkane and add -ene

- for chains of four or more carbon atoms, a number must be added in front of '–ene' to indicate the position of the $C = C$ bond
- the $C = C$ bond position is given by the lower number of the two carbon atoms in it
- the positions of any alkyl branches are indicated using the rules for branched chain alkanes (not needed for Intermediate 2 Chemistry).

For example, pent-2-ene

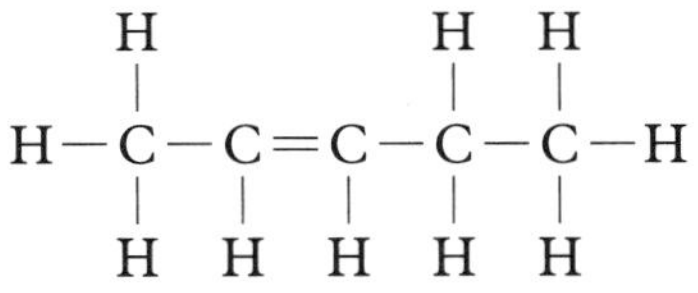

full structural formula

$CH_3CH{=}CHCH_2CH_3$

shortened structural formula

For example, oct-3-ene

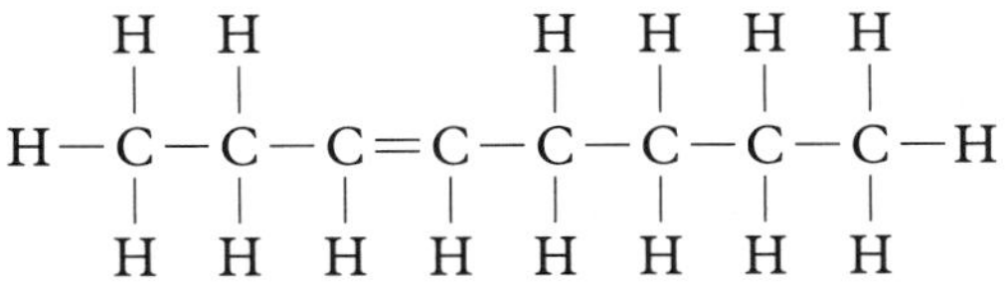

full structural formula

$CH_3CH_2CH{=}CHCH_2CH_2CH_2CH_3$

shortened structural formula

Cycloalkanes

○ The **cycloalkanes**, another subset of the set of hydrocarbons, are similar to the alkanes in that they contain only single $C - C$ bonds. Examples include:

Name	Molecular formula and state at room temperature	Shortened structural formula
Cyclopropane	$C_3H_6(g)$	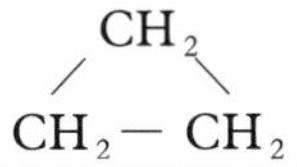
Cyclobutane	$C_4H_8(g)$	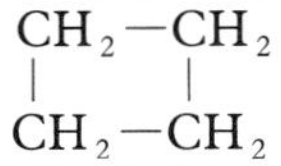
Cyclopentane	$C_5H_{10}(l)$	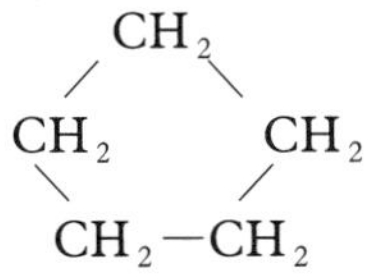

○ Cyclopropane is the simplest cycloalkane because it takes a minimum of three carbon atoms to form a ring structure.
○ The general formula for the cycloalkanes is C_nH_{2n}.
○ Full structural formulae are often used for the cycloalkanes – for example

49

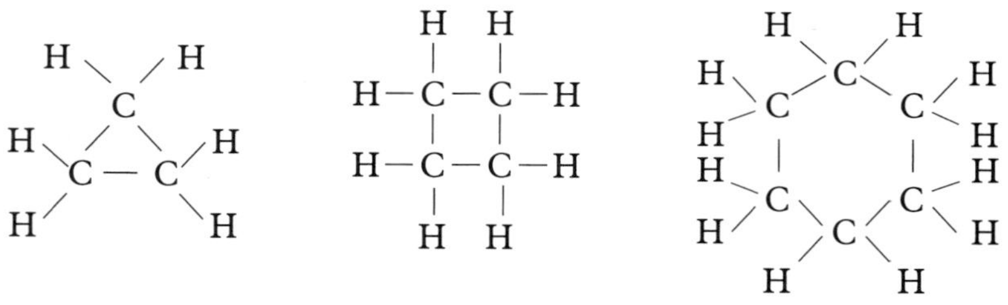

cyclopropane   cyclobutane   cyclohexane

- ○ The rules for naming the cycloalkanes are:
  - – the number of carbon atoms in the ring gives the name of the parent alkane
  - – add the prefix cyclo- in front of the name for the parent alkane.

Homologous series

- ○ The alkanes, alkenes and cycloalkanes are examples of **homologous series**, being groups of compounds which:
  - – can be represented by a general formula
  - – have similar chemical properties
  - – show a gradual change in physical properties, such as boiling points.

- ○ We will meet other homologous series shortly – alkanols and alkanoic acids for example. Every homologous series is characterised by a **functional group** – a group of atoms with characteristic chemical activity:
  - – the alkanols have the hydroxyl (– OH) group
  - – the alkanoic acids have the carboxyl (– COOH) group.

Isomers

- ○ **Isomers** are compounds with the same molecular formula but different structural formulae. For example, there are two isomers with molecular formula C_4H_{10}:

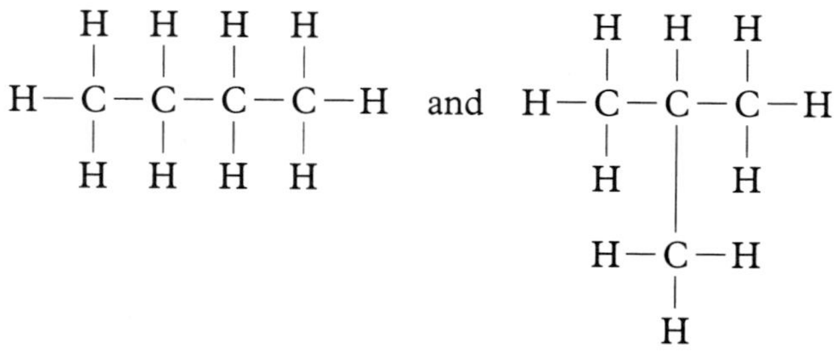

butane   (2-)methylpropane

- ○ For every cycloalkane there is an alkene that is an isomer – for example propene and cyclopropane both have molecular formula C_3H_6:

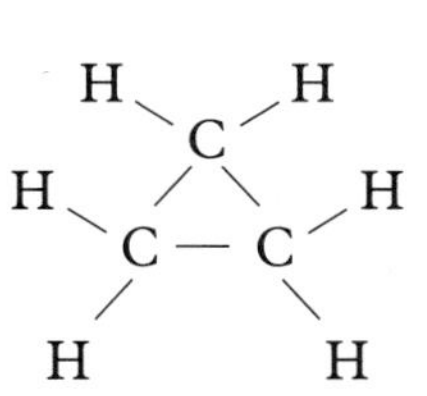

cyclopropane 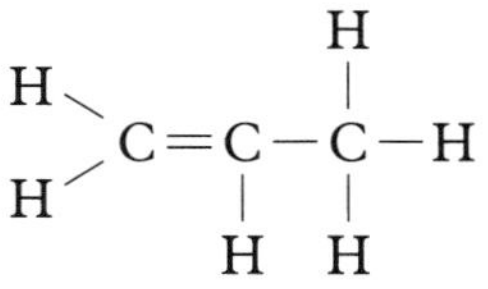   propene

○ There are no isomers for methane (CH_4), ethane (C_2H_6), propane (C_3H_8) and ethene (C_2H_4).

Alkanols

The **alkanols** are a homologous series in which a hydrogen atom in an alkane has been replaced by a **hydroxyl group**, $-OH$. Examples include:

Name	Molecular formula and state at room temperature	Shortened structural formula
Methanol	$CH_3OH(l)$	CH_3OH
Ethanol	$C_2H_5OH(l)$	CH_3CH_2OH
Propan-1-ol	$C_3H_7OH(l)$	$CH_3CH_2CH_2OH$
Propan-2-ol	$C_3H_7OH(l)$	$CH_3CHOHCH_3$

○ The general formula for the alkanols is $C_nH_{2n+1}OH$.
○ Full structural formulae are often used for the alkanols – for example

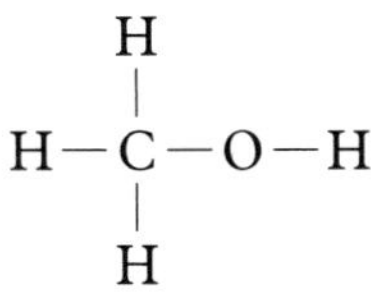

methanol 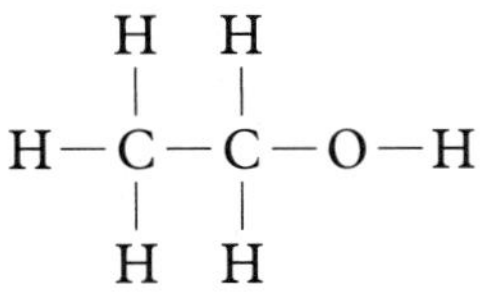   ethanol 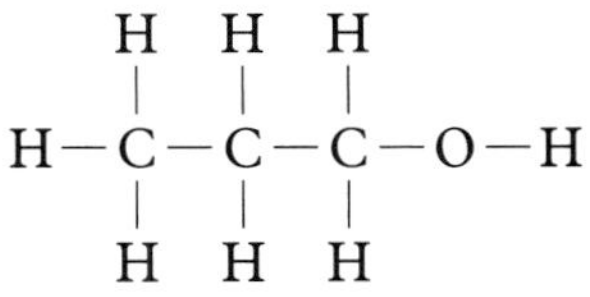   propan-1-ol

○ The rules for naming alkanols are:
  – look for the longest carbon chain containing the $-OH$ group to give the name of the parent alkane
  – remove the '–e' ending of the parent alkane and add -ol
  – for chains of three or more carbon atoms, a number must be added in front of '–ol' to indicate the position of the $-OH$ group
  – the carbon chain is numbered from the end which gives the smaller number to the $-OH$ group
  – the positions of any alkyl branches can be indicated using the rules for branched chain alkanes (not needed for Intermediate 2 Chemistry).

  For example, pentan-3-ol

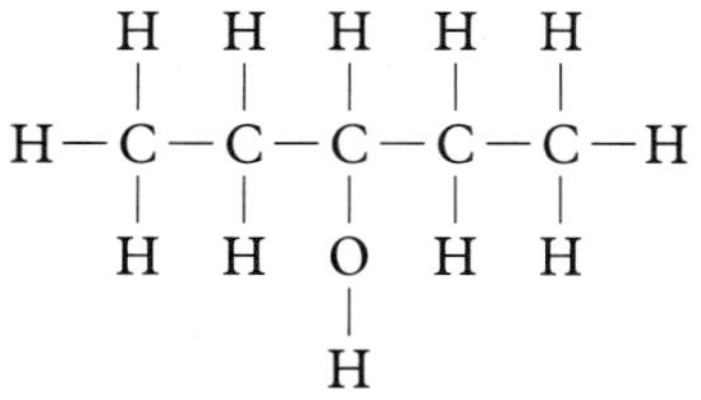

$$H-\overset{\displaystyle H}{\underset{\displaystyle H}{\overset{\displaystyle |}{\underset{\displaystyle |}{C}}}}-\overset{\displaystyle H}{\underset{\displaystyle H}{\overset{\displaystyle |}{\underset{\displaystyle |}{C}}}}-\overset{\displaystyle H}{\underset{\displaystyle O}{\overset{\displaystyle |}{\underset{\displaystyle |}{C}}}}-\overset{\displaystyle H}{\underset{\displaystyle H}{\overset{\displaystyle |}{\underset{\displaystyle |}{C}}}}-\overset{\displaystyle H}{\underset{\displaystyle H}{\overset{\displaystyle |}{\underset{\displaystyle |}{C}}}}-H$$

H — C — C — C — C — C — H $CH_3CH_2CHOHCH_2CH_3$

full structural formula   shortened structural formula

For example, butan-2-ol

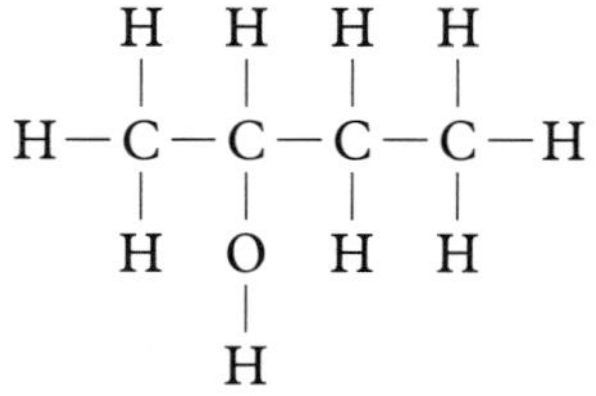

H — C — C — C — C — H $CH_3CHOHCH_2CH_3$

full structural formula   shortened structural formula

Alkanoic acids

○ The **alkanoic acids** are a homologous series in which a $-CH_3$ group in an alkane has been replaced by a **carboxyl group** $-COOH$. Examples include:

Name	Molecular formula and state at room temperature	Shortened structural formula
Methanoic acid	HCOOH(l)	HCOOH
Ethanoic acid	CH_3COOH(l)	CH_3COOH
Propanoic acid	C_2H_5COOH(l)	CH_3CH_2COOH
Butanoic acid	C_3H_7COOH(l)	$CH_3CH_2CH_2COOH$

○ The general formula for the alkanoic acids is $C_nH_{2n+1}COOH$. (Note that in methanoic acid $n = 0$.)

○ Full structural formulae are often used for the alkanoic acids – for example

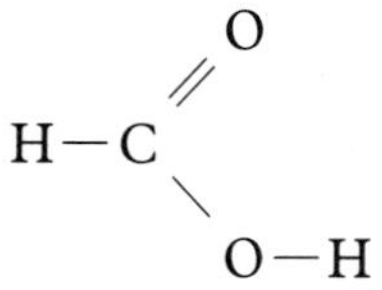

methanoic acid 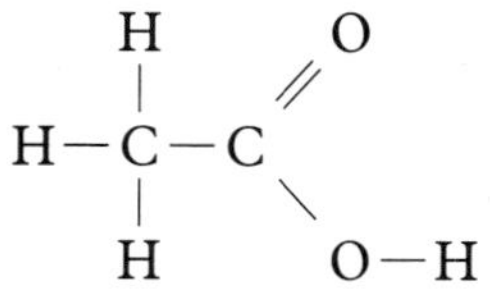   ethanoic acid 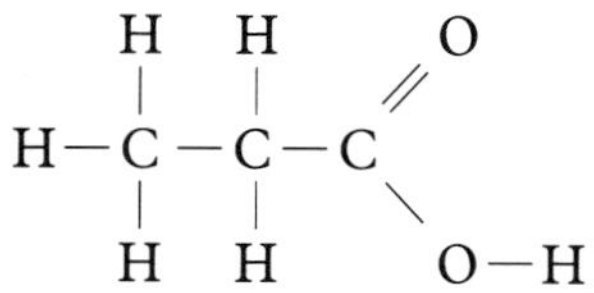   propanoic acid

○ The rules for naming alkanoic acids are:

  – look for the longest carbon chain containing the $-COOH$ group to give the name of the parent alkane

  – remove the '–e' ending from the parent alkane and add -oic acid

  – the positions of alkyl branches can be indicated starting the numbering of the carbon chain at the carbon atom in the $-COOH$ group (not needed for Intermediate 2 Chemistry).

For example, pentanoic acid

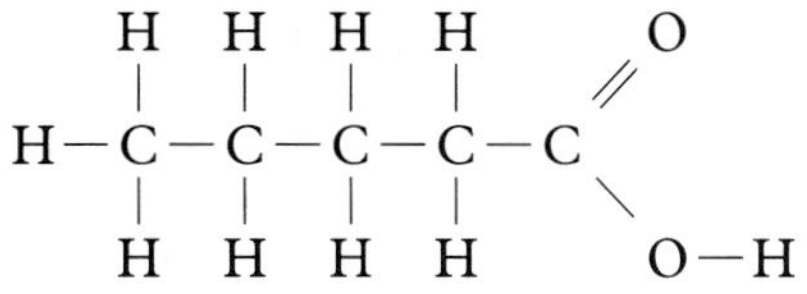

$$H-\underset{\underset{H}{|}}{\overset{\overset{H}{|}}{C}}-\underset{\underset{H}{|}}{\overset{\overset{H}{|}}{C}}-\underset{\underset{H}{|}}{\overset{\overset{H}{|}}{C}}-\underset{\underset{H}{|}}{\overset{\overset{H}{|}}{C}}-C\underset{O-H}{\overset{O}{\diagup\diagdown}}$$

$CH_3CH_2CH_2CH_2COOH$

full structural formula

shortened structural formula

Esters

○ **Esters** are compounds formed in the reaction between an alkanol and an alkanoic acid. They contain the **ester group**,

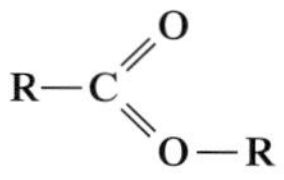

$$R-C\underset{O-R}{\overset{O}{\diagup\diagdown}}$$

'R' represents any alkyl group.

○ Esters are named according to the alkanol and alkanoic acid from which they are formed.

○ Alkanols produce alkyl esters – for example:
  – methanol produces methyl esters
  – ethanol produces ethyl esters
  – propan-1-ol produces propyl esters.

○ Alkanoic acids produce alkanoate esters – for example:
  – methanoic acid produces methanoate esters
  – ethanoic acid produces ethanoate esters
  – propanoic acid produces propanoate esters.

○ An alkanol and an alkanoic acid react to give esters with the name alkyl alkanoate – for example:
  – methanol and ethanoic acid give methyl ethanoate
  – ethanol and propanoic acid give ethyl propanoate.

○ When an alkanol and an alkanoic acid react, water is formed in addition to the ester:

$$\text{alkanol} + \text{alkanoic acid} \rightarrow \text{ester} + \text{water}$$

○ In equations, the alkanoic acid molecule is normally written first and then the alkanol molecule is written 'back-to-front' – for example

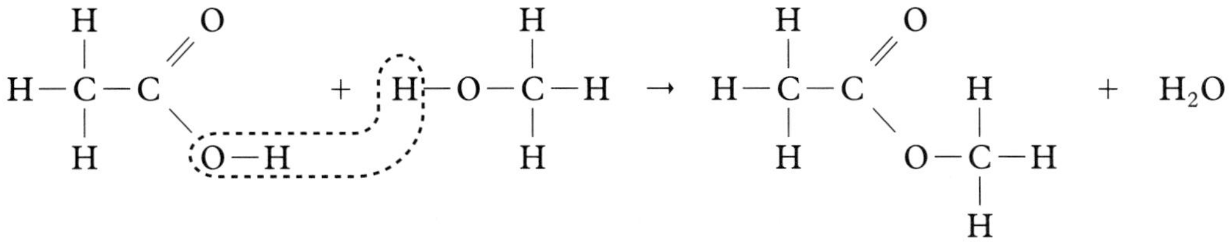

ethanoic acid   methanol   methyl ethanoate

Using shortened structural formulae:

$$CH_3COOH + C_2H_5OH \rightarrow CH_3COOCH_3 + H_2O$$

53

Using propanoic acid and ethanol the reaction is

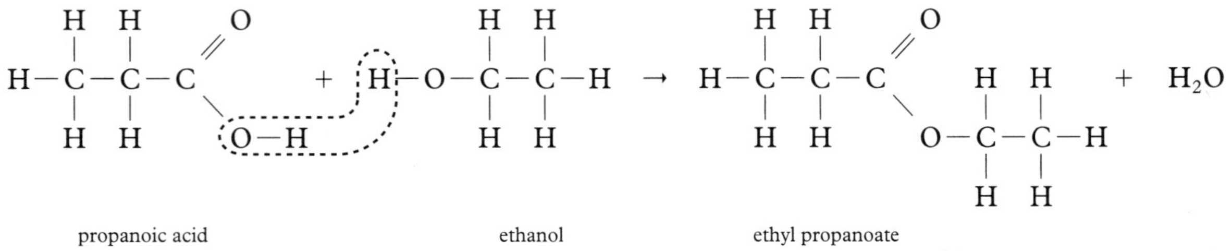

propanoic acid   ethanol   ethyl propanoate

Using shortened structural formulae:

$$C_2H_5COOH + C_2H_5OH \rightarrow C_2H_5COOC_2H_5 + H_2O$$

○ Esters can be broken down into the corresponding alkanol and alkanoic acid – for example ethyl methanoate can be broken down into ethanol and methanoic acid.

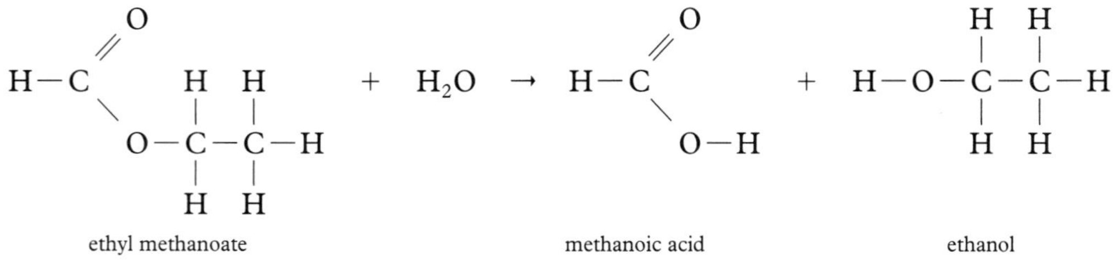

ethyl methanoate   methanoic acid   ethanol

Using shortened structural formulae:

$$HCOOC_2H_5 + H_2O \rightarrow HCOOH + C_2H_5OH$$

QUESTIONS

1 The grid contains full structural formulae for some hydrocarbons.

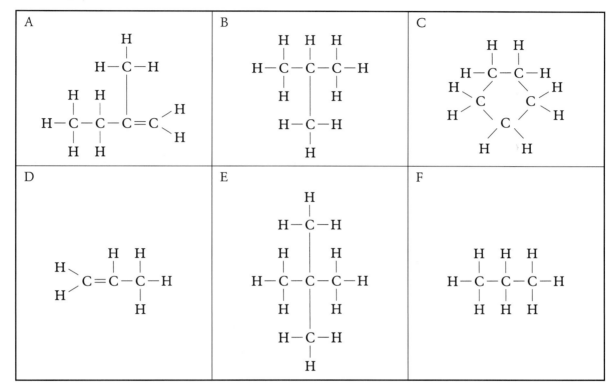

a) Which *two* hydrocarbons are isomers?
b) Which hydrocarbon is an isomer of butane?
c) Which hydrocarbon is an isomer of the simplest cycloalkane?
d) Which hydrocarbon is a branched chain alkene?
e) Which hydrocarbon is 2-methylpropane?

2 a) The general formula for the alkanes is C_nH_{2n+2}. What is the general formula for both the alkenes and the cycloalkanes?
b) Use the general formula for the alkanes to give molecular formulae for the alkanes which have
i) 22 carbon atoms in each molecule
ii) 28 hydrogen atoms in each molecule.

3 Draw full structural formulae for each of the following:
a) heptane
b) 2,4-dimethylhexane
c) 3-ethyl-2-methylpentane

4 Give systematic names for each of the following:
a) $CH_3CH_2CH(CH_3)CH_2CH_3$
b) $CH_3CH(CH_3)CH(CH_3)CH(CH_3)$
$CH_2CH_2CH_3$
c)

$$CH_2$$
$$CH_2 \qquad CH_2$$
$$CH_2 \qquad CH_2$$
$$CH_2$$

5 Draw full structural formulae for each of the following:
a) pent-3-ene
b) oct-2-ene
c) hept-1-ene

6 Give systematic names for each of the following:
   a) $CH_2 = CHCH_3$
   b) $CH_3CH_2CH_2CH = CH_2$
   c) $CH_3CH_2CH_2CH = CHCH_2CH_2CH_3$

7 Draw shortened structural formulae for each of the following:
   a) ethene
   b) but-2-ene
   c) hex-3-ene

8 All of the compounds in the grid contain the elements carbon, hydrogen and oxygen.

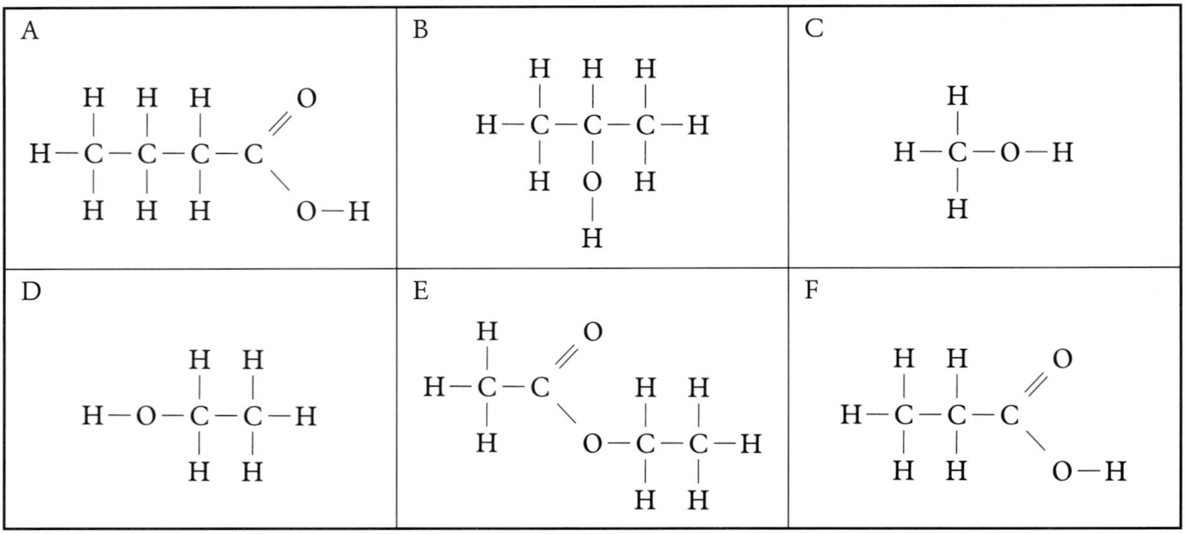

A	B	C

H H H   O
│ │ │   ⫽
H—C—C—C—C
│ │ │   \
H H H   O—H
	H H H
	│ │ │
	H—C—C—C—H
	│ │ │
	H O H
	   │
	   H
		
		    H
		    │
		H—C—O—H
		    │
		    H

D	E	F

   H H
   │ │
H—O—C—C—H
   │ │
   H H
	H O
	│ ⫽
	H—C—C H H
	│ \ │ │
	H O—C—C—H
	        │ │
	        H H
		H H O
		│ │ ⫽
		H—C—C—C
		│ │ \
		H H O—H

a) Identify the *two* compounds which are isomers.
b) Identify the ester.
c) Identify the compound which would react with an alkanoic acid to give an ethyl ester.
d) Identify the compound which would react with an alkanol to give a propanoate ester.

9 Draw full structural formulae for each of the following:
   a) butan-2-ol
   b) heptan-1-ol
   c) hexan-3-ol

10 Give systematic names for each of the following:
   a) CH_3CH_2OH
   b) $CH_3CH_2CH_2CHOHCH_2CH_3$
   c) $CH_2OHCH_2CH_2CH_3$

11 Draw full structural formulae for each of the following:
   a) ethanoic acid
   b) butanoic acid
   c) hexanoic acid

12 Give systematic names for each of the following:
   a) HCOOH
   b) C_2H_5COOH
   c) $CH_3CH_2CH_2CH_2COOH$

13 Give full structural formulae for each of the following:
   a) methyl methanoate
   b) ethyl propanoate
   c) propyl ethanoate

14 Give systematic names for each of the following:
  a) CH_3COOCH_3
  b) $CH_3CH_2CH_2COOCH_2CH_3$
  c)

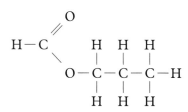

15 Give the name for the ester which is formed on reaction between each of the following:
  a) ethanol and ethanoic acid
  b) methanol and pentanoic acid
  c) butan-1-ol and propanoic acid

16 Give the names for the alkanol and the alkanoic acid which react to produce each of the following esters:
  a) methyl propanoate
  b) ethyl hexanoate
  c) propyl ethanoate

17 Which of the following compounds has an isomer?

A
$$H-\overset{\overset{\displaystyle H}{|}}{\underset{\underset{\displaystyle H}{|}}{C}}-\overset{\overset{\displaystyle H}{|}}{\underset{\underset{\displaystyle H}{|}}{C}}-H$$

B
$$\overset{\overset{\displaystyle H}{|}}{\underset{\underset{\displaystyle H}{|}}{C}}=\overset{\overset{\displaystyle H}{|}}{\underset{\underset{\displaystyle H}{|}}{C}}$$

C
$$H-\overset{\overset{\displaystyle H}{|}}{\underset{\underset{\displaystyle H}{|}}{C}}-\overset{\overset{\displaystyle H}{|}}{\underset{\underset{\displaystyle H}{|}}{C}}-\overset{\overset{\displaystyle H}{|}}{\underset{\underset{\displaystyle H}{|}}{C}}-H$$

D
$$\overset{\overset{\displaystyle H}{|}}{\underset{\underset{\displaystyle H}{|}}{C}}=\overset{\overset{\displaystyle H}{|}}{C}-\overset{\overset{\displaystyle H}{|}}{\underset{\underset{\displaystyle H}{|}}{C}}-H$$

18 Which of the following is a structural formula for methyl ethanoate?

A
$$CH_3-\overset{\overset{\displaystyle O}{\|}}{C}-O-CH_3$$

B
$$CH_3-\overset{\overset{\displaystyle O}{\|}}{C}-O-CH_2-CH_3$$

C
$$CH_3-CH_2-\overset{\overset{\displaystyle O}{\|}}{C}-O-CH_3$$

D
$$H-\overset{\overset{\displaystyle O}{\|}}{C}-O-CH_2-CH_3$$

(2.3) Reactions of carbon compounds

Addition reactions

○ The alkanes and cycloalkanes are said to be **saturated hydrocarbons** because they contain only carbon-to-carbon single covalent bonds, C–C.

○ The alkenes are **unsaturated hydrocarbons** because they contain at least one carbon-to-carbon double covalent bond, C=C.

○ The presence of the C=C bond makes alkenes more reactive than alkanes and cycloalkanes.

○ An **addition reaction** is one in which two or more molecules join to produce a single larger molecule and nothing else.

○ The alkenes undergo many addition reactions due to the presence of the C=C bond. Three examples are given below.

alkene + hydrogen → alkane (heat + catalyst needed) – for example:
ethene   + hydrogen   →   ethane

$$\underset{H}{\overset{H}{\diagdown}}C=C\underset{H}{\overset{H}{\diagup}} \quad + \quad H-H \quad \rightarrow \quad H-\underset{\underset{H}{|}}{\overset{\overset{H}{|}}{C}}-\underset{\underset{H}{|}}{\overset{\overset{H}{|}}{C}}-H$$

alkene + water → alkanol (heat + catalyst needed) – for example:
ethene   + water   →   ethanol

$$\underset{H}{\overset{H}{\diagdown}}C=C\underset{H}{\overset{H}{\diagup}} \quad + \quad H-O\diagdown_{H} \quad \rightarrow \quad H-\underset{\underset{H}{|}}{\overset{\overset{H}{|}}{C}}-\underset{\underset{O}{|}}{\overset{\overset{H}{|}}{C}}-H$$
$$\underset{\underset{H}{|}}{}$$

alkene + bromine → dibromoalkane – for example:
ethene   + bromine   → 1,2-dibromoethane

$$\underset{H}{\overset{H}{\diagdown}}C=C\underset{H}{\overset{H}{\diagup}} \quad + \quad Br-Br \quad \rightarrow \quad H-\underset{\underset{Br}{|}}{\overset{\overset{H}{|}}{C}}-\underset{\underset{Br}{|}}{\overset{\overset{H}{|}}{C}}-H$$

○ Because it takes place readily and is accompanied by a clear colour change, the reaction with bromine is used as a *test for unsaturation*.

○ Alkenes decolourise bromine solution rapidly in the test for unsaturation, whereas alkanes and cycloalkanes do not. For example:

but-2-ene + bromine → addition compound
colourless *orange/red* *colourless*

$$\underset{\begin{array}{c}|\\ H\end{array}}{H}-\overset{\begin{array}{c}H\\ |\end{array}}{C}-\overset{\begin{array}{c}H\\ |\end{array}}{C}=\overset{\begin{array}{c}H\\ |\end{array}}{C}-\overset{\begin{array}{c}H\\ |\end{array}}{\underset{\begin{array}{c}|\\ H\end{array}}{C}}-H \quad + \quad Br-Br \quad \rightarrow \quad H-\overset{\begin{array}{c}H\\ |\end{array}}{\underset{\begin{array}{c}|\\ H\end{array}}{C}}-\overset{\begin{array}{c}H\\ |\end{array}}{\underset{\begin{array}{c}|\\ Br\end{array}}{C}}-\overset{\begin{array}{c}H\\ |\end{array}}{\underset{\begin{array}{c}|\\ Br\end{array}}{C}}-\overset{\begin{array}{c}H\\ |\end{array}}{\underset{\begin{array}{c}|\\ H\end{array}}{C}}-H$$

Bromine solution can appear yellow when very dilute.

Prescribed Practical Activity

Testing for unsaturation

Safety note

Because of the toxic nature of bromine, this experiment is best carried out in a fume chamber

- The aim of this experiment is to test several liquid hydrocarbons for unsaturation using bromine solution and, based on the results, to draw possible structural formulae for them.
- Add a small volume of a liquid hydrocarbon to a test tube.
- Add a few drops of bromine solution to the hydrocarbon.
- Carefully shake the contents of the test tube and record your observations.
- Decolourisation indicates the presence of at least one $C=C$ bond.
- No decolourisation indicates that there are no $C=C$ bonds present. All carbon-to-carbon covalent bonds present are single, $C-C$.

- If a hydrocarbon with molecular formula C_5H_{10} *did* decolourise bromine solution, it could be pent-1-ene, the full structural formula of which is

$$\overset{H}{\underset{H}{>}}C=\overset{\begin{array}{c}H\\ |\end{array}}{C}-\overset{\begin{array}{c}H\\ |\end{array}}{\underset{\begin{array}{c}|\\ H\end{array}}{C}}-\overset{\begin{array}{c}H\\ |\end{array}}{\underset{\begin{array}{c}|\\ H\end{array}}{C}}-\overset{\begin{array}{c}H\\ |\end{array}}{\underset{\begin{array}{c}|\\ H\end{array}}{C}}-H$$

- If a hydrocarbon with molecular formula C_5H_{10} did *not* decolourise bromine solution then it could be cyclopentane, the full structural formula of which is

$$\begin{array}{c}H\qquad\quad H\\ \diagdown\qquad\diagup\\ H\qquad C\qquad H\\ \diagdown\quad\diagup\ \diagdown\quad\diagup\\ C\qquad\qquad C\\ H\diagup\ \diagdown\qquad\diagup\ \diagdown H\\ H-C-C-H\\ |\qquad|\\ H\qquad H\end{array}$$

Cracking

○ Fractional distillation of crude oil gives more long-chain hydrocarbons (mainly alkanes) than are useful for industrial purposes.

○ **Cracking** is a method of producing smaller, more useful molecules by heating large hydrocarbon molecules in the presence of a catalyst.

○ Cracking can be carried out in the laboratory using an aluminium oxide or silicate catalyst.

○ Cracking alkane molecules usually produces a mixture of smaller alkanes and alkenes. For example, decane can be cracked to give a mixture of octane and ethene:

$$C_{10}H_{22} \rightarrow C_8H_{18} + C_2H_4$$

○ There are insufficient hydrogen atoms in an alkane molecule for it to produce only smaller alkanes on cracking, so some unsaturated molecules are produced as well.

○ The presence of a catalyst allows cracking to take place at a lower temperature, thus saving energy and making it more economical.

○ During cracking the catalyst becomes coated with carbon, which reduces its efficiency, but this is easily burned off to regenerate it.

Prescribed Practical Activity

Cracking

Safety note

Because of the toxic nature of bromine, this experiment is best carried out in a fume chamber. Avoid 'suck-back'.

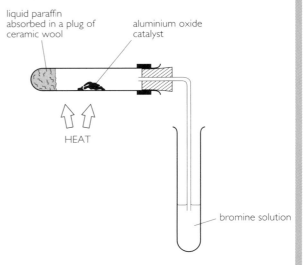
liquid paraffin absorbed in a plug of ceramic wool

aluminium oxide catalyst

HEAT

bromine solution

- The aim of the experiment is to crack liquid paraffin, which is a mixture of long chain alkanes, and to show that some of the products are unsaturated.

- The apparatus below is set up and then the catalyst is heated strongly for about 10 seconds.

- The flame is then directed very briefly at the *nearest* end of the plug for a second or two to vaporise some paraffin, then back to the catalyst again.

- The process of alternate heating is repeated as required in order to maintain a flow of 'cracked gas'.

- When the bromine solution has been decolourised, a special procedure *MUST* be followed when heating is to stop.

- The delivery tube must be removed from the solution *before* heating stops so as to

prevent 'suck-back'. 'Suck-back' happens when pressure inside the apparatus drops and air pressure forces the cold solution up the delivery tube and into the still very hot test tube.

- One constituent of liquid paraffin is eicosane, $C_{20}H_{42}$. When cracked, one possible pair of products is a liquid called heptadecane, $C_{17}H_{36}$, and the unsaturated gas propene, C_3H_6.

$$C_{20}H_{42} \rightarrow C_{17}H_{36} + C_3H_6$$

Ethanol

○ Ethanol, for alcoholic drinks, can be made by **fermentation** of glucose derived from a fruit or vegetable:

$$\text{glucose} \rightarrow \text{ethanol} + \text{carbon dioxide}$$
$$C_6H_{12}O_6 \rightarrow 2C_2H_5OH + 2CO_2$$

○ An enzyme in yeast acts as a catalyst for the reaction.

○ In the laboratory, ethanol can be made by adding yeast to a solution of glucose in water, or to a fruit juice, keeping the mixture warm for a few days, and then distilling off the ethanol.

○ There is a limit to the ethanol concentration of fermented drinks because yeast cells die if the concentration rises above about 12%.

○ Because water boils at 100 °C and ethanol boils at 78 °C, the two can be separated by **distillation**. This enables 'spirits' (high ethanol content drinks such as whisky) to be produced.

○ Alcoholic drinks, if taken in excess, can have damaging effects to health and mind.

○ Ethanol is a widely used industrial chemical and fermentation cannot meet the market demand for it.

○ Industrial ethanol is manufactured by the **catalytic hydration** of ethane – heat and a catalyst are needed:

$$\text{ethene} + \text{water} \rightarrow \text{ethanol}$$

$$
\begin{array}{ccc}
\underset{H}{\overset{H}{>}}C=C\underset{H}{\overset{H}{<}} & + & H-O\underset{H}{} & \rightarrow & H-\underset{\underset{H}{|}}{\overset{\overset{H}{|}}{C}}-\underset{\underset{\underset{\underset{H}{|}}{O}}{|}}{\overset{\overset{H}{|}}{C}}-H
\end{array}
$$

○ Ethanol can be dehydrated to give ethene by passing the vapour over hot aluminium oxide which acts as a catalyst. This reaction, which is given by many alkanols, can be carried out in the laboratory (see Figure 1).

Figure 1 Ethanol can be dehydrated to give ethene

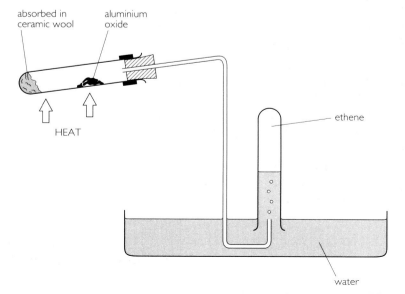

61

ethanol → ethene + water

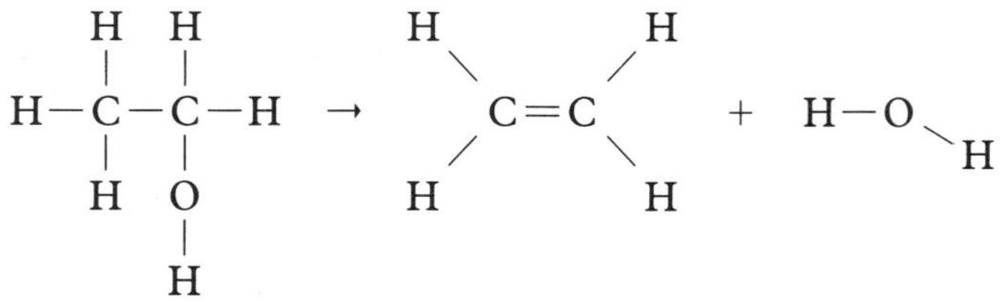

$$\begin{array}{l} \overset{\displaystyle H}{\underset{\displaystyle H}{\overset{\displaystyle |}{\underset{\displaystyle |}{C}}}} \end{array}$$

ethanol → ethene + water

H–C–C–H → C=C + H–O

○ Ethanol, mixed with petrol, can be used as a fuel for cars.

○ The ethanol for this purpose is obtained from sugar cane by fermentation and distillation. Sugar cane is a **renewable energy resource**.

Making and breaking esters

○ The reaction between an alkanol and an alkanoic acid to produce an ester and water is a **reversible reaction**. The alkanol and alkanoic acid react to form the ester and water, but at the same time ester and water react to reform the alkanol and alkanoic acid.

alkanol + alkanoic acid ⇌ ester + water

○ Esters can be made by mixing small equal volumes of an alkanol and an alkanoic acid in a test tube, adding a few drops of concentrated sulphuric acid as a catalyst and warming the mixture in a water bath. The ester may be separated from unreacted alkanol and alkanoic acid by pouring the final mixture into water in a beaker. The insoluble ester floats on the surface.

○ Simple esters are sweet-smelling liquids and are widely used as fruit flavourings – for example pear drops. In the cosmetics industry they are used in perfumes and nail varnish remover. Industrially they are important solvents.

○ The formation of an ester from an alkanol and an alkanoic acid is an example of a **condensation reaction** – one in which two molecules join to form a single larger molecule, with water formed at the same time. The hydroxyl functional group of the alkanol joins with the carboxyl functional group of the alkanoic acid to form the ester link and water – for example

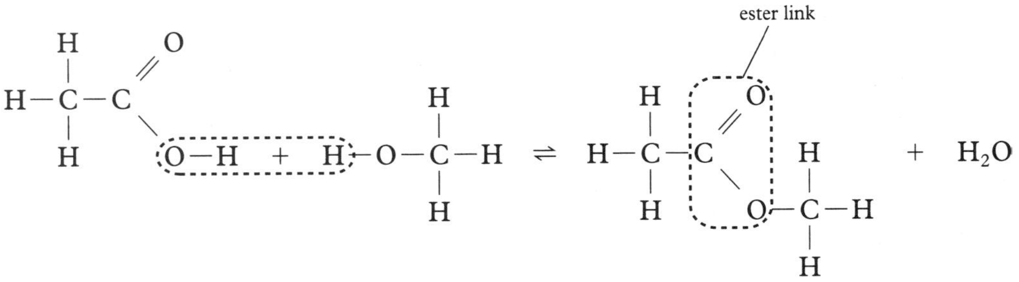

ester link

H–C–C   O–H + H–O–C–H ⇌ H–C–C   H + H_2O

ethanoic acid   +   methanol   ⇌   methyl ethanoate   +   water

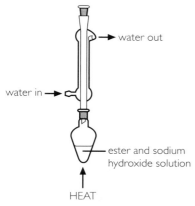

water out

water in →

ester and sodium
hydroxide solution

HEAT

Figure 2 Apparatus for reflux

❍ The breaking down of an ester into an alkanol and an alkanoic acid by reaction with water is an example of a **hydrolysis reaction**.

❍ A hydrolysis reaction is one in which a large molecule is broken down into smaller molecules by reaction with water.

❍ In order to hydrolyse an ester, it is refluxed with sodium hydroxide solution as catalyst, rather than sulphuric acid. The reaction goes to completion with sodium hydroxide because the alkali catalyst reacts with the alkanoic acid as it is formed, preventing the reverse reaction from taking place. The apparatus for reflux is shown in Figure 2. The alkanol can be distilled off and the alkanoic acid reformed by the addition of hydrochloric acid.

❍ The forward reaction in ester formation involves condensation, whereas the reverse reaction involves hydrolysis

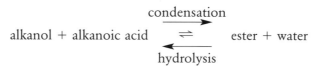

condensation

alkanol + alkanoic acid ⇌ ester + water

hydrolysis

QUESTIONS

1 Look at these different different types of reactions.

A $C_3H_6 + Br_2 \rightarrow C_3H_6Br_2$
B Water + ethyl ethanoate → ethanol + ethanoic acid
C $C_{13}H_{28} \rightarrow C_4H_8 + C_9H_{20}$
D $CH_3COOH + CH_3OH \rightarrow CH_3COOCH_3 + H_2O$
E $C_2H_4 + H_2O \rightarrow C_2H_5OH$
F $C_6H_{12}O_6 \rightarrow 2C_2H_5OH + 2CO_2$

Identify:
a) the addition reaction(s)
b) the fermentation reaction
c) the catalytic hydration reaction
d) the condensation reaction
e) the hydrolysis reaction
f) the cracking reaction

2 A hydrocarbon with molecular formula C_5H_{10} decolourises bromine solution rapidly. Based on this information it is likely to be
A a cycloalkane
B an alkane
C an alkene
D a saturated hydrocarbon

3 Dehydration of a simple alkanol involves
A removal of hydrogen and formation of an alkanoic acid
B removal of hydrogen and formation of an alkene
C removal of water and formation of an alkanoic acid
D removal of water and formation of an alkene

4 a) Explain the meaning of the terms *saturated hydrocarbon* and *unsaturated hydrocarbon*.
b) Describe the test for unsaturation.

5 a) What is produced when hydrogen adds on to an alkene?
b) Copy and complete:
i) propene + hydrogen → _____
ii) _____ + hydrogen → ethane

6 Draw full structural formulae for the molecules produced when bromine adds on to
a) propene
b) pent-2-ene

7 a) Explain the meaning of the word *cracking*.
  b) Why is cracking an important reaction industrially?
  c) Which metal oxide is used as a catalyst for cracking?
  d) How is the carbon coating which forms on this catalyst removed?
  e) Catalysts speed up the reaction rate, but there is another purpose for the catalyst used during cracking. State this purpose.

8 Given that the molecules cracked in the following examples produce only *two* products, copy and complete the equations:
  a) $C_9H_{20} \rightarrow C_3H_6 + -$
  b) $C_2H_6 \rightarrow C_2H_4 + -$

9 Example 8(b) is unusual because cracking usually produces a mixture of smaller alkanes and alkenes. Why is it not possible that a mixture of smaller alkanes alone is formed?

10 Ethanol has, for many years, been produced by fermentation.
  a) What catalyst is used in this process?
  b) Why is there a limit (of about 12%) to the ethanol concentration that can be produced by fermentation?
  c) How can alcoholic drinks with a higher ethanol concentration be produced?

11 Ethanol is in such wide use that fermentation alone cannot meet the demand.
  a) From which simple hydrocarbon molecule is ethanol made?
  b) What name is given to this process?
  c) How can this process be reversed?

12 Ester formation involves a reversible reaction.
  a) What is meant by the term *reversible*?
  b) From which two types of compounds are esters made?
  c) How could you make a simple ester and separate it from the final mixture?

13 Simple esters are pleasant-smelling liquids.
  a) Give three uses of simple liquid esters.
  b) What type of reaction is involved in ester formation?

14 a) What is meant by the term *functional group*?
  b) Copy the reaction equation below, and then circle and label the ester, carboxyl and hydroxyl functional groups.

2.4 **Plastics and synthetic fibres**

○ Synthetic materials are made by the chemical industry.
○ Most plastics and synthetic fibres are made from chemicals derived from crude oil.
○ Examples of plastics include polythene, polystyrene, perspex, PVC, bakelite, formica, nylons and silicones.
○ Kevlar, which is very strong, and poly(ethenol), which dissolves in water readily, are recently developed plastics.
○ Examples of synthetic fibres include nylons and polyesters. Terylene is a polyester fibre.
○ For some uses, synthetic materials have advantages over natural materials, and vice versa.

Synthetic and natural materials compared

○ Natural fibres have some advantages – for example:
  – wool is warm and soft
  – cotton is cool and soft.
○ Synthetic fibres also have some advantages – for example:
  – water evaporates quickly from polyester and nylon
  – nomex (similar to kevlar) is flame-resistant and is used to make fire-proof clothing.
○ Plastics are often used in place of traditional materials such as iron, steel, wood, leather and paper.
○ The particular uses of plastics are related to their properties.
○ Most plastics have low density, are good heat and electrical insulators and are water-resistant.
○ PVC has replaced iron for pipes, for example, because it is durable and water-resistant.
○ Kevlar has replaced wood in hockey sticks and other materials in tennis racquets because it is strong and flexible.

Plastic	Property	Use
Kevlar	Strong and flexible	Bullet-proof clothing, hockey sticks
Poly(ethenol)	Water soluble	Soluble laundry bags in hospitals
Poly(tetrafluoroethene)	Non-stick	Lining in cooking pots and pans
Formica	Heat resistant	Kitchen work surfaces
Nylon	Low friction	Curtain rails

Plastics and pollution

○ Most plastics are not **biodegradable** – they do not rot away naturally. Because of their low density and durability they can cause environmental problems as plastic litter.

○ Natural materials – such as wood, paper and cardboard – are biodegradable.

○ Biopol is a recently developed biodegradable plastic.

○ Plastics can – and should – be recycled.

○ Like all carbon-containing fuels, when plastics are burned they can give off poisonous carbon monoxide gas if combustion is incomplete.

○ Some plastics can give off other **toxic** gases when they burn or smoulder.

○ The gases given off during burning or smouldering can be related to the elements present in the plastic.

Gas produced	Elements present	Example of plastic
Carbon monoxide, CO	C	Polystyrene
Hydrogen chloride, HCl	H and Cl	PVC
Hydrogen cyanide, HCN	H, C and N	Polyurethane

Thermoplastics and thermosetting plastics

○ A **thermoplastic** is one which softens on heating and can be reshaped – for example polythene, polystyrene, PVC, perspex and nylon.

○ A **thermosetting plastic** does not soften on heating and cannot be reshaped – for example bakelite and formica.

○ Whether a plastic is thermosetting or thermoplastic can be investigated by touching it with a hot iron nail.

Addition polymerisation

○ Plastics and synthetic fibres are made up of long chain molecules called **polymers** – very large molecules formed by the joining of many small molecules called **monomers**.

○ Monomers are small molecules that can join together to form a large polymer molecule.

○ **Polymerisation** is the process whereby many small monomer molecules join to form a large polymer molecule.

○ **Addition polymerisation** is a process in which many small monomer molecules join to form one large polymer molecule *and nothing else* – for example the formation of poly(ethene) from ethene.

○ The monomers that undergo addition polymerisation are unsaturated – they contain a carbon-to-carbon double covalent bond. They join together by the opening of the carbon-to-carbon double bond.

○ Many addition polymers are made from small unsaturated molecules produced by cracking – for example ethene and propene.

○ Some monomers are made from ethene – for example chloroethene (vinyl chloride) and phenylethene (styrene).

○ A polymer name is obtained by putting 'poly' in front of the bracketed monomer name:

Monomer name	Polymer name
Ethene	Poly(ethene)
Propene	Poly(propene)
Chloroethene	Poly(chloroethene)
Phenylethene	Poly(phenylethene)

○ No brackets are used in the traditional polymer names – for example:

  – chloroethene is also known as vinyl chloride, giving the traditional polymer name of polyvinyl chloride or PVC

  – phenylethene is also known as styrene, giving the traditional polymer name of polystyrene.

○ The repeating unit, or the structure, of an addition polymer can be drawn given the monomer structure, and vice versa. Examples include:

ethene monomers   poly(ethene) polymer   repeating unit

propene monomers   poly(propene) polymer   repeating unit

phenylethene monomers   poly(phenylethene) polymer   repeating unit
(styrene)   (polystyrene)

Condensation polymerisation

○ **Condensation polymerisation** is a process in which many small monomer molecules join to form one large polymer molecule with water, or some other small molecule, formed at the same time.

○ Condensation polymers are made from monomers with *two* functional groups per molecule.

Polyesters

○ **Polyesters** are condensation polymers made by reacting a dicarboxylic acid with a diol.

○ A dicarboxylic acid has two carboxyl groups in each molecule – for example terephthalic acid, $HOOCC_6H_4COOH$.

○ A diol has two hydroxyl groups in each molecule – for example ethane-1,2-diol, $HOCH_2CH_2OH$.

○ Terephthalic acid and ethane-1,2-diol undergo condensation polymerisation to produce the polyester known as Terylene.

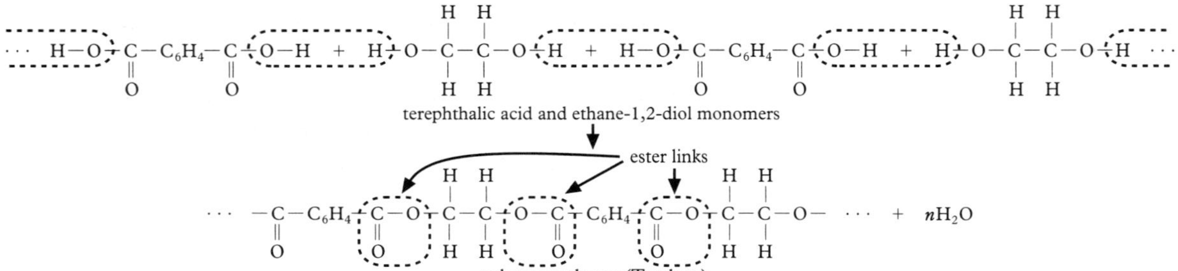

terephthalic acid and ethane-1,2-diol monomers

polyester polymer (Terylene)

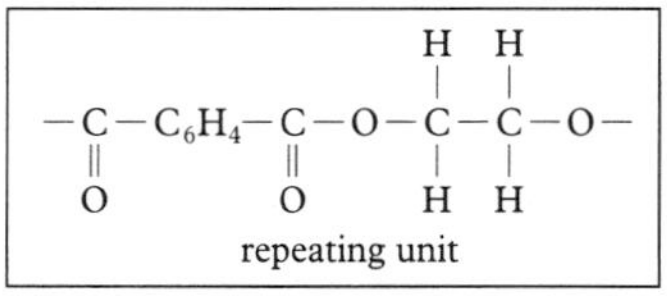

repeating unit

○ As with addition polymers, the repeating unit, or the structure, of a condensation polymer can be drawn given the monomer structures, and vice versa.

○ In order to obtain the structures of the monomers, given the structure of a polyester polymer, the following procedure can be followed:

– locate the region of the ester link

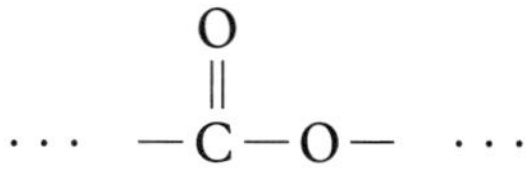

$$\cdots \ -\overset{\overset{\textstyle O}{\|}}{C}-O- \ \cdots$$

– water adds on like this

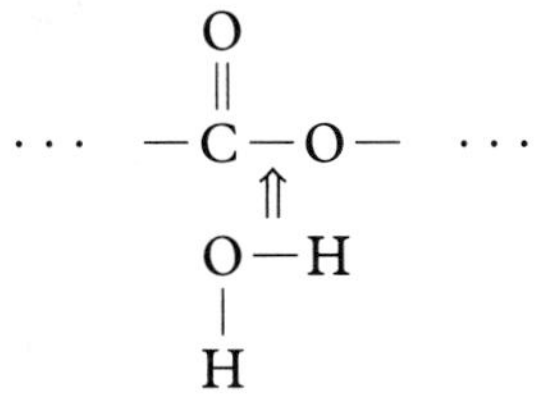

$$
\cdots \ \ -\overset{\overset{\textstyle O}{\|}}{C}-O- \ \ \cdots
$$

$$
\Uparrow
$$

$$
\overset{\textstyle O-H}{\underset{\textstyle H}{|}}
$$

– the carboxyl and hydroxyl groups in the monomers can now be reformed

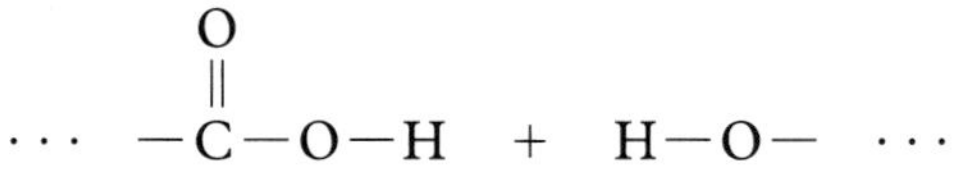

$$
\cdots \ \ -\overset{\overset{\textstyle O}{\|}}{C}-O-H \ \ + \ \ H-O- \ \ \cdots
$$

Polyamides

○ **Polyamides** are condensation polymers usually made by reacting a dicarboxylic acid with a diamine.

○ An **amine** is a compound in which a hydrogen atom in an alkane, or other hydrocarbon, has been replaced by the amine (or amino) functional group $-NH_2$.

○ **Diamines** contain two $-NH_2$ groups in each molecule.

○ Polyamides are held together by the **amide link** $-CO-NH-$, the structural formula of which is

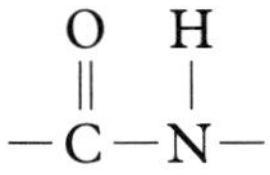

$$
-\overset{\overset{\textstyle O}{\|}}{C}-\overset{\overset{\textstyle H}{|}}{N}-
$$

○ Polyamides are more commonly called 'nylons', since the word nylon refers to a group of synthetic condensation polymers each of which contains amide links.

○ The polyamide nylon-6,6 is made from hexane-1,6-dioic acid and 1,6-diaminohexane:

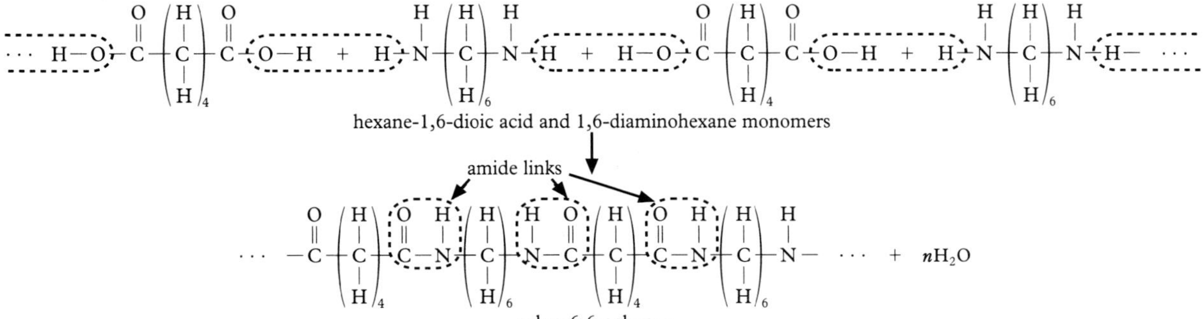

hexane-1,6-dioic acid and 1,6-diaminohexane monomers

amide links

nylon-6,6 polymer $+ \ nH_2O$

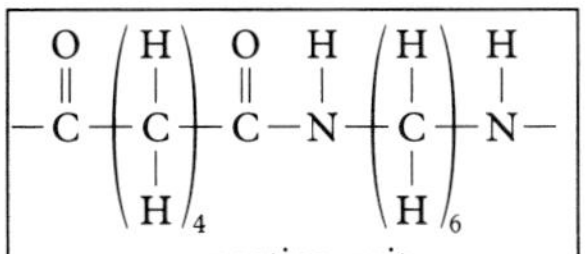

repeating unit

○ The repeating unit, or the structure, of a polyamide can be drawn given the monomer structures, or vice versa.

○ In order to obtain the structures of the monomers, given the structure of the polyamide polymer, the following procedure can be followed–

– locate the region of the amide link

$$
\cdots \quad -\overset{\overset{\displaystyle O}{\|}}{C}-\overset{\overset{\displaystyle H}{|}}{N}- \quad \cdots
$$

– water adds on like this

$$
\cdots \quad -\overset{\overset{\displaystyle O}{\|}}{C}-\overset{\overset{\displaystyle H}{|}}{\underset{\underset{\displaystyle H}{|}}{\underset{O-H}{\Uparrow}}}N- \quad \cdots
$$

– the carboxyl and **amine groups** in the monomers can now be reformed

$$
\cdots \quad -\overset{\overset{\displaystyle O}{\|}}{C}-O-H \quad + \quad H-\overset{\overset{\displaystyle H}{|}}{N}- \quad \cdots
$$

QUESTIONS

1 The grid contains carbon-based molecules, some of which are capable of polymerising.

A	B	C
$\begin{array}{c} H\ \ H\ \ H \\ \|\ \ \ \|\ \ \ \| \\ H-C-C-C-O-H \\ \|\ \ \ \|\ \ \ \| \\ H\ \ H\ \ H \end{array}$	$\begin{array}{c} H \\ \diagdown \\ \ \ \ N \\ H \diagup \end{array} \left(\begin{array}{c} H \\ \| \\ C \\ \| \\ H \end{array}\right)_6 \begin{array}{c} H \\ N \diagup \\ \diagdown H \end{array}$	$\begin{array}{c} H\ \ H \\ \|\ \ \ \| \\ H-C-C=C \\ \|\ \ \ \ \ \diagdown \\ H \end{array} \begin{array}{c} H \\ \\ H \end{array}$
D	E	F
$\begin{array}{c} O\ \ \ \ \ \ \ \ \ \ \ \ \ \ O \\ \diagdown\ \ \ \ \ \ \ \ \ \ \ \ \diagup \\ C-C_6H_4-C \\ \diagup\ \ \ \ \ \ \ \ \ \ \ \ \diagdown \\ H-O\ \ \ \ \ \ \ \ \ O-H \end{array}$	$\begin{array}{c} O\ \ \ \ \ \ H\ \ H \\ \diagdown\ \ \ \ \ \ \ \|\ \ \ \| \\ C-C-C-H \\ \diagup\ \ \ \ \|\ \ \ \| \\ H-O\ \ \ H\ \ H \end{array}$	$\begin{array}{c} H\ \ H \\ \|\ \ \ \| \\ H-O-C-C-O-H \\ \|\ \ \ \| \\ H\ \ H \end{array}$

a) Which molecule(s) could undergo addition polymerisation?

b) Which *two* molecules could react to form a polyamide (nylon)?

c) Which *two* molecules could react to form a polyester?

d) Which molecule(s) could not take part in a polymerisation reaction?

2 For which of the following would you *not* use a thermoplastic?
A A curtain hook
B A frying pan handle
C An ice cube tray
D A milk crate

3 Which of the following is part of a condensation polymer?

A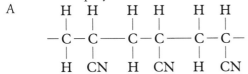

B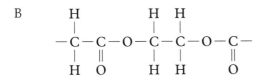

C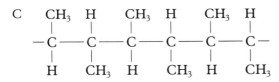

D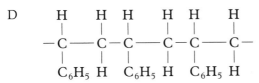

4 Which of the following plastics dissolves in water?
A Kevlar
B Terylene
C Polythene
D Poly(ethenol)

5 Which of the following is made from a monomer which is unsaturated and therefore decolourises bromine solution?
A Polyester
B Polypropene
C Polyamide
D Kevlar

6 From which of the following are most plastics and synthetic fibres made?
A Crude oil
B Natural gas
C Wood
D Coal

7 Bakelite and urea-methanal are thermosetting plastics which are widely used for electrical sockets and plugs.
a) Explain the meaning of the term *thermosetting plastic*.
b) Explain why a thermoplastic, such as polythene, is considered to be unsuitable for such applications.

8 Give *two* reasons for each of the following:
a) Most people prefer wool to other fibres for a jumper or cardigan
b) Cotton is the preferred material for a summer T-shirt
c) PVC has largely replaced iron for gutter pipes
d) Kevlar is used in tennis racquet frames.

9 State the meaning of the terms:
a) monomer
b) polymer
c) addition polymer
d) condensation polymer

10 PVC is derived from the old name 'polyvinyl chloride', but its modern name is poly(chloroethene) which has the following structure:

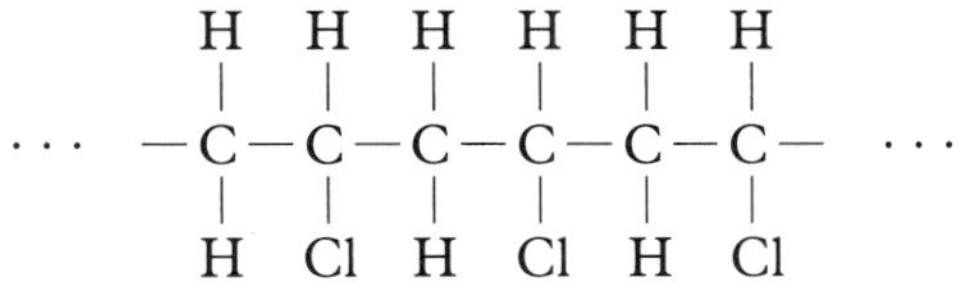

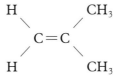

a) Give the modern name for the monomer from which PVC is made.
b) Draw the repeating unit in PVC.
c) Draw the full structural formula of the monomer from which PVC is made.

11 2-methylpropene has the following structure:

$$
\begin{array}{ccc}
H & & CH_3 \\
\diagdown & & \diagup \\
& C=C & \\
\diagup & & \diagdown \\
H & & CH_3
\end{array}
$$

a) Draw a section of the polymer which is produced when 2-methylpropene polymerises, showing three monomer molecules joined together.
b) Draw the repeating unit for this polymer.

12 One of several kinds of nylon used throughout the world is nylon-6,10. Part of the structure for this nylon is as follows:

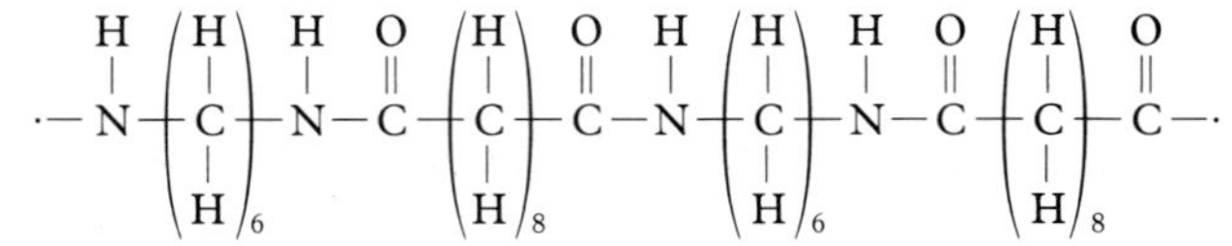

$$-N\underset{H}{\overset{H}{\underset{|}{\overset{|}{\vert}}}}-\left(\underset{H}{\overset{H}{\underset{|}{\overset{|}{C}}}}\right)_6-N-\overset{O}{\overset{\Vert}{C}}-\left(\underset{H}{\overset{H}{\underset{|}{\overset{|}{C}}}}\right)_8-\overset{O}{\overset{\Vert}{C}}-N-\left(\underset{H}{\overset{H}{\underset{|}{\overset{|}{C}}}}\right)_6-N-\overset{O}{\overset{\Vert}{C}}-\left(\underset{H}{\overset{H}{\underset{|}{\overset{|}{C}}}}\right)_8-\overset{O}{\overset{\Vert}{C}}-\cdot$$

a) Draw the structure of the amide link.
b) Draw the repeating unit in nylon-6,10.
c) Draw structural formulae for the monomers from which this nylon is made.

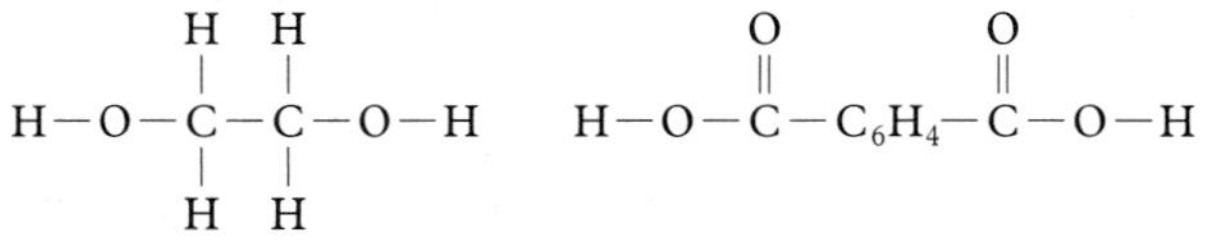

$$H-O-\underset{H}{\overset{H}{\underset{|}{\overset{|}{C}}}}-\underset{H}{\overset{H}{\underset{|}{\overset{|}{C}}}}-O-H \qquad H-O-\overset{O}{\overset{\Vert}{C}}-C_6H_4-\overset{O}{\overset{\Vert}{C}}-O-H$$

13 Terylene is made from the two monomers shown below:
a) What type of polymer is formed when these two monomers polymerise?
b) Draw part of the polymer showing four monomer molecules joined.

14 Nylon is a polymer. The monomers are shown below are used to produce a nylon.
a) Draw a section of the polymer showing the three monomer units linked together.
b) What feature of their structure makes these molecules suitable for use as monomers?

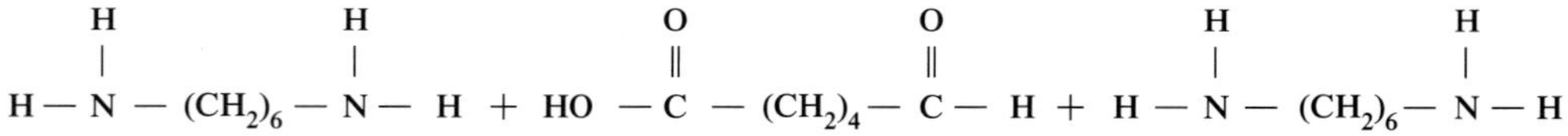

$$H-\overset{H}{\overset{|}{N}}-(CH_2)_6-\overset{H}{\overset{|}{N}}-H \;+\; HO-\overset{O}{\overset{\Vert}{C}}-(CH_2)_4-\overset{O}{\overset{\Vert}{C}}-H \;+\; H-\overset{H}{\overset{|}{N}}-(CH_2)_6-\overset{H}{\overset{|}{N}}-H$$

Carbohydrates

○ **Carbohydrates** form an important class of food made by plants.

○ The first stage in a plant's production of carbohydrates is **photosynthesis**. This produces simple carbohydrates, such as glucose, from carbon dioxide and water using the Sun's energy in the presence of chlorophyll. Oxygen is also released in this process.

○ Foods with a high carbohydrate content include bread, sugar, honey, rice, potatoes and pasta.

○ Carbohydrates supply the body with *energy*, the simpler ones combining with oxygen in exothermic reactions producing carbon dioxide and water.

○ Laboratory experiments can be carried out to show that carbohydrates burn and that they release a lot of energy on combustion.

○ All carbohydrates are covalent compounds containing the elements carbon, hydrogen and oxygen, with two hydrogen atoms for every one oxygen atom.

○ Adding concentrated sulphuric acid (a dehydrating agent) to a carbohydrate removes the hydrogen and oxygen atoms to leave a black mass of carbon.

Carbohydrates large and small

○ The carbohydrates which we use as food can be divided into **sugars** and **starch**.

○ Sugar molecules are fairly small – examples of sugars include glucose, fructose, maltose and sucrose (table sugar).

○ Glucose and fructose both have the molecular formula $C_6H_{12}O_6$ – they are called **monosaccharides**.

○ Maltose and sucrose have molecular formula $C_{12}H_{22}O_{11}$ – they are called **disaccharides**.

○ Starch is a natural condensation polymer made of many glucose molecules joined together – starch is called a **polysaccharide**.

○ The molecular formula for starch is $(C_6H_{10}O_5)_n$ where 'n' is a large number.

○ Plants convert glucose into starch for storing energy. The process is an example of condensation polymerisation.

$$nC_6H_{12}O_6 \rightarrow (C_6H_{10}O_5)_n + nH_2O$$

many glucose molecules → one starch molecule + water

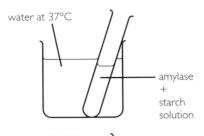

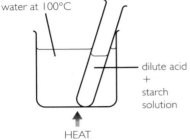

Figure 1 Two ways of hydrolysing starch

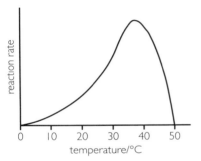

Figure 2 The effect of temperature on the rate of a body enxyme-catalysed reaction

Chemical tests for carbohydrates

◯ Most sugars (including glucose, fructose and maltose) produce an orange-red colour when heated with **Benedict's** (or Fehling's) **solution**. Sucrose does *not* give this result.

◯ Starch can be distinguished from other carbohydrates by the addition of **iodine solution**, with which it produces a dark blue colour.

Breaking down carbohydrates

◯ During digestion, starch is hydrolysed to glucose.

$$(C_6H_{10}O_5)_n + nH_2O \rightarrow nC_6H_{12}O_6$$

◯ In the body, the enzyme **amylase** catalyses the hydrolysis of starch to glucose.

◯ Glucose can pass through the gut wall but starch cannot. Glucose is carried by the bloodstream to body cells where **respiration** occurs, releasing energy

$$C_6H_{12}O_6 + 6O_2 \rightarrow 6CO_2 + 6H_2O$$

◯ Starch can be hydrolysed rapidly using amylase at about body temperature (37 °C) or more slowly using a dilute acid at 100 °C as shown in Figure 1.

◯ Body enzymes such as amylase work best at body temperature (their optimum temperature). At higher temperatures they are destroyed. The effect of temperature on a reaction catalysed by a body enzyme is shown in Figure 2.

Prescribed Practical Activity

Hydrolysis of starch

• The aim of this experiment is to hydrolyse starch in the presence of either an enzyme or an acid and to demonstrate that the enzyme or acid catalyses the reaction.

• The enzyme-catalysed reaction is the most straightforward.

• Apparatus as shown in Figure 1 above can be used, but a test tube containing starch solution only should also be present as a control.

• After just a few minutes, the enzyme (amylase) catalyses sufficient hydrolysis of the starch for a positive test with Benedict's solution to be obtained by raising the temperature of the water in the beaker.

• The control does not give a positive test with Benedict's solution, proving that it must be the enzyme that caused the hydrolysis reaction – it was acting as a catalyst.

• In this reaction, starch is hydrolysed to give a sugar which gives a positive test with Benedict's solution.

Proteins

- ○ **Proteins** form an important class of food made by plants.
- ○ Foods with a high protein content include fish, meat, cheese, eggs, peas, beans and lentils.
- ○ Proteins are the major structural materials of animal tissue and are involved in the maintenance and regulation of life processes. They include haemoglobin (the red pigment and oxygen carrier in blood), enzymes and many hormones – for example insulin.
- ○ Proteins are condensation polymers made of many **amino acid** molecules joined together.
- ○ Amino acids contain two functional groups – the amine group $-NH_2$ and the carboxyl group $-COOH$. The amino acids used to make proteins have the following structure, where 'R' stands for the rest of the molecule:

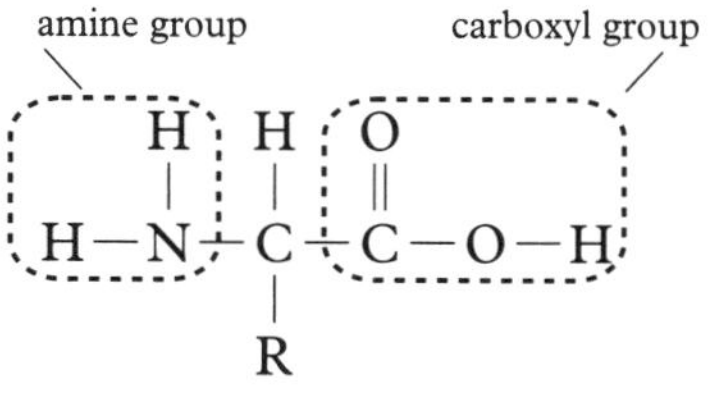

amine group   carboxyl group

$$
\begin{array}{ccc}
H & H & O \\
| & | & \| \\
H-N-C-C-O-H \\
& | \\
& R
\end{array}
$$

- ○ There are more than twenty amino acids that can make up proteins, some examples are:

Name	R group	Full structural formula
Glycine	$- H$	$\begin{array}{ccc} H & H & O \\ | & | & \| \\ H-N-C-C-O-H \\ & | \\ & H \end{array}$
Alanine	$\begin{array}{c} H \\ | \\ -C-H \\ | \\ H \end{array}$	$\begin{array}{ccc} H & H & O \\ | & | & \| \\ H-N-C-C-O-H \\ & | \\ & H-C-H \\ & | \\ & H \end{array}$
Serine	$\begin{array}{c} H \\ | \\ -C-O-H \\ | \\ H \end{array}$	$\begin{array}{ccc} H & H & O \\ | & | & \| \\ H-N-C-C-O-H \\ & | \\ & H-C-H \\ & | \\ & O \\ & | \\ & H \end{array}$

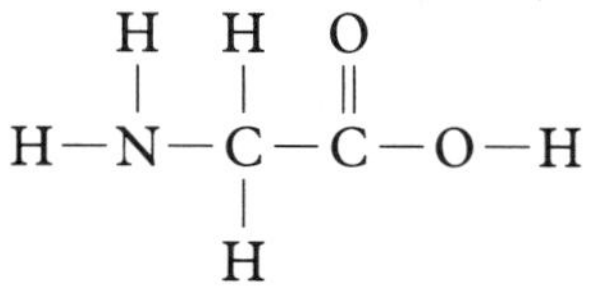

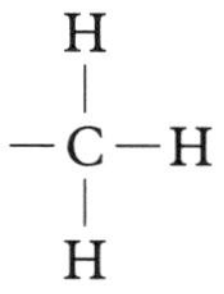

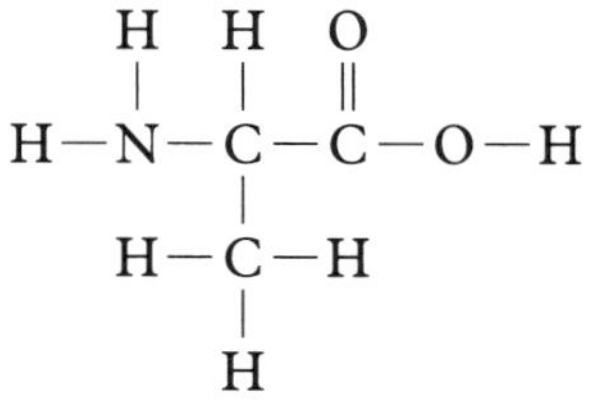

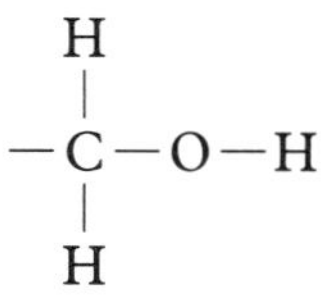

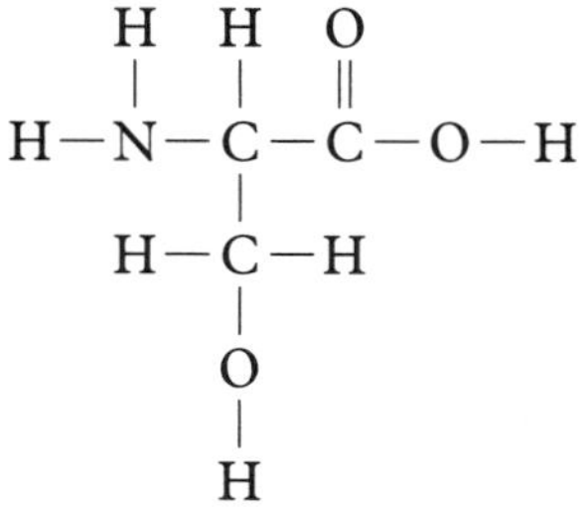

- ○ As with nylon (polyamide) formation, protein formation is due to a reaction between the amine and carboxyl groups.

○ The condensation polymerisation of amino acids during protein formation produces the same link, $-CO-NH-$, that is present in polyamides. In polyamides it is called the amide link, but in proteins it is called the **peptide link** – for example

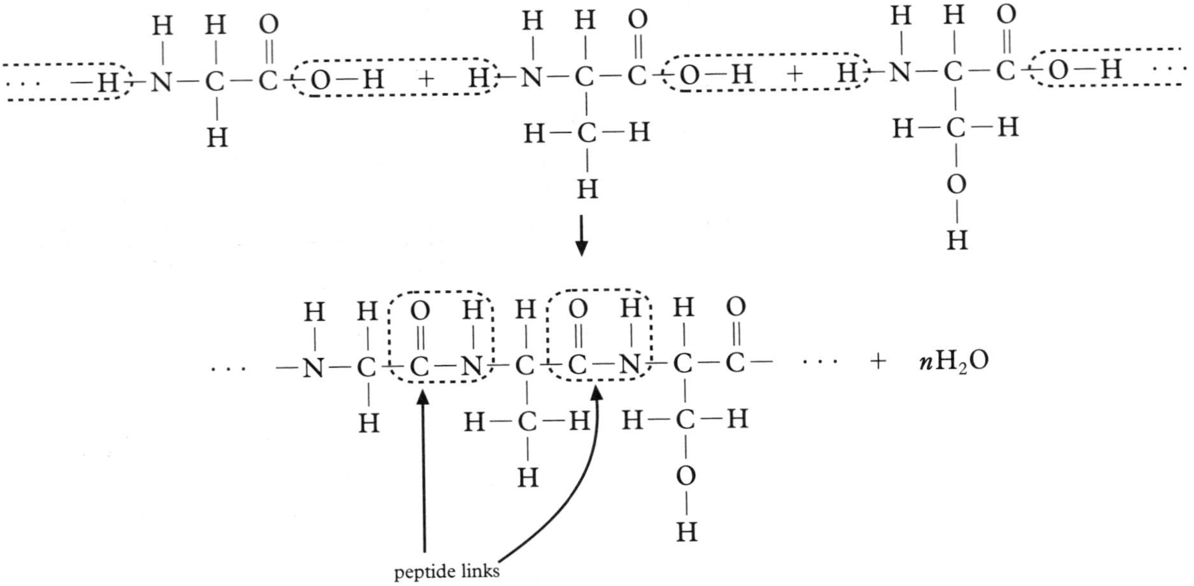

$$
\cdots -H \!\!\rightarrow\!\! N-C-C\!\!\leftarrow\!\!O-H \ + \ H\!\!\rightarrow\!\! N-C-C\!\!\leftarrow\!\!O-H \ + \ H\!\!\rightarrow\!\! N-C-C\!\!\leftarrow\!\!O-H \ \cdots
$$

$$
\downarrow
$$

$$
\cdots -N-C-C-N-C-C-N-C-C- \ \cdots \ + \ n H_2O
$$

peptide links

○ During digestion, enzyme hydrolysis of dietary protein produces amino acids. These are carried by the blood to be built up into proteins specific to the body's needs.

○ In the laboratory, proteins can be hydrolysed to produce amino acids by refluxing with hydrochloric acid.

○ The structural formulae of amino acids formed from the hydrolysis of proteins can be identified from the structure of a section of the protein by the following procedure:

– locate the region of the peptide link

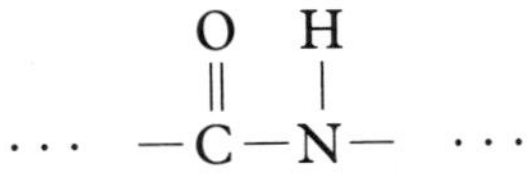

$$
\cdots \ -C-N- \ \cdots
$$

– water adds on like this

$$
\cdots \ -C-N- \ \cdots
$$
$$
O-H
$$
$$
H
$$

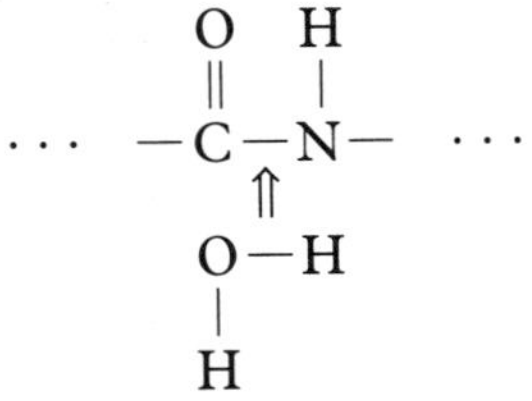

– the carboxyl and amine groups in the amino acids can now be reformed.

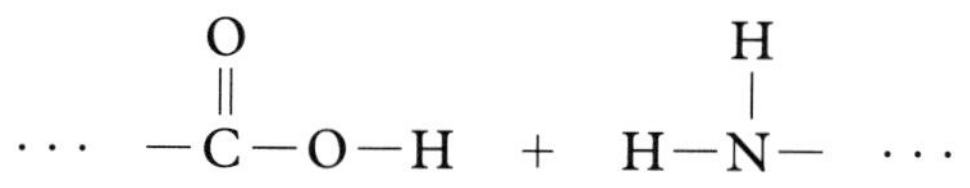

$$\cdots \ -\overset{\overset{\textstyle O}{\|}}{C}-O-H \ + \ H-\overset{\overset{\textstyle H}{|}}{N}- \ \cdots$$

Fats and oils

❍ Natural **fats and oils** can be classified according to their origin as animal, vegetable or marine. Examples include:

Animal	Vegetable	Marine
Beef fat	Olive oil	Cod liver oil
Chicken fat	Sunflower oil	Halibut oil

❍ Foods with a high fat or oil content include butter, margarine, cheese, cream and nuts.

❍ The lower melting points of oils compared to those of fats, of a similar molecular size, is related to the higher degree of **unsaturation** of oil molecules. The oil molecules contain more carbon-to-carbon double covalent bonds, C=C.

❍ In the test for unsaturation using bromine solution, oils decolourise more bromine solution than an equal mass of fat.

❍ The conversion of oils into hardened fats, as in the making of margarine from sunflower oil, involves the partial removal of unsaturation by hydrogenation (the addition of hydrogen) – heat and a catalyst are needed.

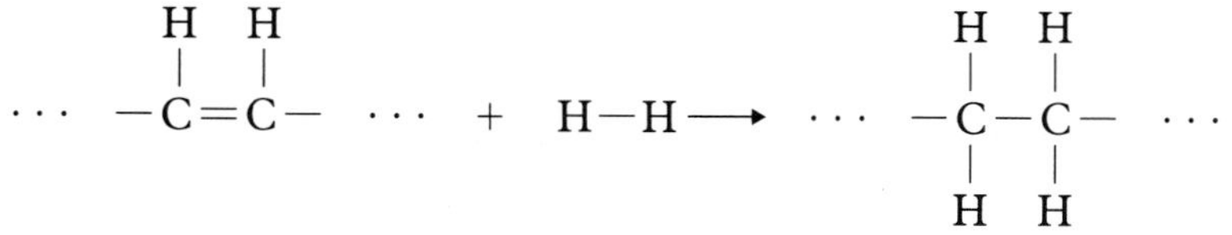

$$\cdots \ -\overset{\overset{\textstyle H}{|}}{C}=\overset{\overset{\textstyle H}{|}}{C}- \ \cdots \ + \ H-H \longrightarrow \ \cdots \ -\overset{\overset{\textstyle H}{|}}{\underset{\underset{\textstyle H}{|}}{C}}-\overset{\overset{\textstyle H}{|}}{\underset{\underset{\textstyle H}{|}}{C}}- \ \cdots \quad \text{(heat and catalyst needed)}$$

❍ Fats and oils in our diet supply the body with energy. They are a more concentrated source of energy than carbohydrates – fats and oils produce about twice as much energy as an equal mass of carbohydrate.

❍ Like carbohydrates, fats and oils contain only the elements carbon, hydrogen and oxygen.

❍ Fats and oils are *esters*.

❍ Hydrolysis of oils and fats produces **fatty acids** and glycerol in the ratio of *three* moles of fatty acids to *one* mole of glycerol.

❍ Glycerol is a triol because each molecule has three – OH groups attached to it:

$$
\begin{array}{ccc}
CH_2OH & & H-\overset{\displaystyle H}{\underset{\displaystyle |}{\overset{\displaystyle |}{C}}}-O-H \\[2pt]
| & & | \\[2pt]
CHOH & \text{or} & H-C-O-H \\[2pt]
| & & | \\[2pt]
CH_2OH & & H-\underset{\displaystyle |}{\overset{\displaystyle |}{C}}-O-H \\[2pt]
& & H
\end{array}
$$

The systematic name for glycerol is propane-1,2,3-triol.

- Fatty acids are saturated or unsaturated straight chain carboxylic acids, usually with long chains of carbon atoms. Examples include:

Name	Molecular formula	Structural formula
Stearic acid	$C_{17}H_{35}COOH$	$CH_3(CH_2)_{16}COOH$
Oleic acid	$C_{17}H_{33}COOH$	$CH_3(CH_2)_7CH{=}CH(CH_2)_7COOH$

- Oil and fat molecules have the following structure:

$$
\begin{array}{l}
CH_2OOCR^1 \\
| \\
CHOOCR^2 \\
| \\
CH_2OOCR^3
\end{array}
$$

The hydrocarbon groups (R) may be the same, but are usually different, since three different fatty acid molecules often join with a glycerol molecule to make an oil or fat molecule.

- In the relatively rare case where only one fatty acid is present in an oil or fat molecule, the 3:1 ratio of fatty acid molecules to glycerol molecules produced on hydrolysis of the oil or fat can be seen from the balanced equation – for example

$$
\begin{array}{llll}
CH_2OOCC_{17}H_{35} & & CH_2OH & \\
| & & | & \\
CHOOCC_{17}H_{35} \;+\; 3H_2O \;\rightarrow\; & & CHOH \;+\; & 3C_{17}H_{35}COOH \\
| & & | & \\
CH_2OOCC_{17}H_{35} & & CH_2OH & \\
\text{glycerol stearate} & & \text{glycerol} & \text{stearic acid}
\end{array}
$$

Glycerol stearate is found in animal fat.

- There is some evidence to suggest that there is a link between high intake of saturated fat in the diet and heart disease.

QUESTIONS

1 The grid contains compounds which are natural products.

A	B	C
$\begin{array}{c} H \\ | \\ H{\diagdown} \quad H{-}C{-}C{\diagup}{\diagup}^{O} \\ N{-}C{-}C \\ H{\diagup} \quad | \quad {\diagdown}O{-}H \\ H{-}C{-}H \\ | \\ H \end{array}$	$C_{15}H_{31}COOH$	$C_{12}H_{22}O_{11}$
D	E	F
$C_6H_{12}C_6$	$\begin{array}{c} H \\ | \\ H{-}C{-}O{-}H \\ | \\ H{-}C{-}O{-}H \\ | \\ H{-}C{-}O{-}H \\ | \\ H \end{array}$	$\begin{array}{c} CH_2OOCC_{15}H_{31} \\ | \\ CHOOCC_{17}H_{35} \\ | \\ CH_2OOCC_{17}H_{33} \end{array}$

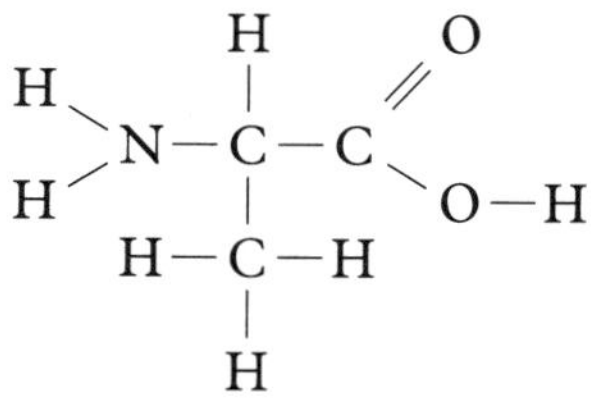

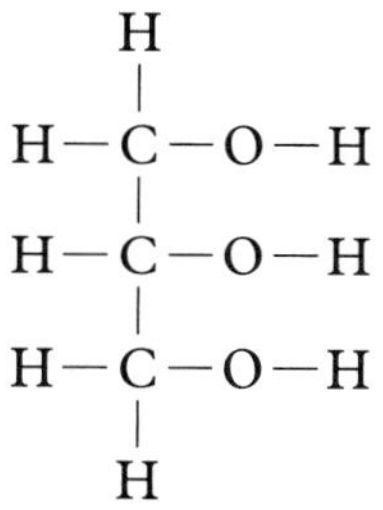

Identify:
a) the disaccharide
b) the fatty acid
c) the amino acid
d) the ester
e) the monosaccharide
f) glycerol

2 Two tests were carried out on a solution which contained two carbohydrates:
I There was a dark blue colour formed on adding iodine solution.
II Heating with Benedict's solution gave an orange-red colour.
The carbohydrates could have been:
A sucrose and starch
B maltose and starch
C glucose and sucrose
D fructose and maltose

3 Which of the following is a condensation polymer?
A glycerol stearate (a fat)
B maltose
C sucrose
D starch

4 Look at these processes:
A polymerisation
B hydrolysis
C respiration
D addition
Pair them up with the following reactions:
a) glucose + oxygen → carbon dioxide + water
b) unsaturated oil + hydrogen → saturated fat
c) glucose → starch + water
d) protein + water → amino acids

5 Give details of chemical tests which could be carried out to distinguish between the following pairs of carbohydrates:
a) sucrose and fructose
b) sucrose and starch

6 a) State the meaning of the word *isomers*.
b) Name two carbohydrates which are isomers having the molecular formula $C_6H_{12}O_6$.
c) Name two carbohydrates which are isomers having the molecular formula $C_{12}H_{22}O_{11}$.

7 When a carbohydrate is burned in oxygen, the products are carbon dioxide and water. Which elements are *proved* to be present in the carbohydrate as a result of this test?

8 a) Write balanced equations for the hydrolysis of **(i)** starch to glucose and **(ii)** sucrose to glucose and fructose.

  b) What type of substances, known as 'biological catalysts', catalyse reactions in the body.

9 The structure of a protein is shown below.

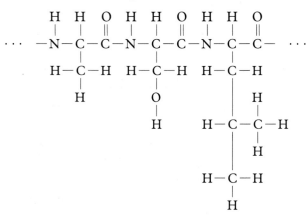

  a) Draw the structure of the peptide (amide) link.

  b) Draw full structural formulae for the amino acids present in the section of protein.

10 a) Explain why each of the following is an important food for humans:
  (i) carbohydrates, **(ii)** proteins, **(iii)** fats.

  b) State which elements are present in each of these foods.

11 Oils have a lower melting point than fats of a similar molecular size.

  a) To which structural feature of the molecules is this related?

  b) Describe a chemical test which would confirm the presence of this structural feature.

12 a) Explain the meaning of the term *fatty acid*.

  b) What is the ratio of moles of fatty acids to moles of glycerol on hydrolysis of a fat or oil?

  c) Give chemical formulae for all of the products which are formed when the fat below is hydrolysed.

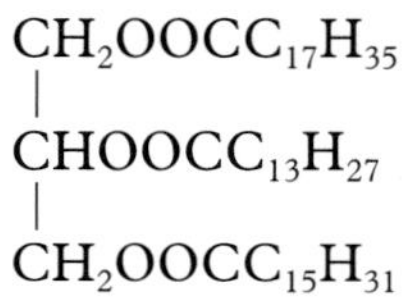

$$CH_2OOCC_{17}H_{35}$$
$$|$$
$$CHOOCC_{13}H_{27}$$
$$|$$
$$CH_2OOCC_{15}H_{31}$$

(2.6) Glossary

2.1

catalytic converter   Part of the exhaust system of a car where platinum and other catalysts change the pollutants carbon monoxide, oxides of nitrogen and unburned hydrocarbons into less harmful carbon dioxide, nitrogen and water vapour.

combustion   A chemical reaction in which a substance combines with oxygen releasing energy in the form of heat and light.

exothermic reaction   One in which heat energy is given out to the surroundings causing an increase in temperature.

flammability   A description of how easy it is to set a substance on fire.

fraction   A group of hydrocarbons with boiling points within a given range obtained from crude oil by fractional distillation.

fractional distillation   A means of separating crude oil into groups of hydrocarbons with similar boiling points (fractions).

fuel   A substance which burns to produce energy.

hydrocarbon   A compound which contains carbon and hydrogen only.

lime water   Calcium hydroxide solution – it is used to test for carbon dioxide, which turns it milky white.

vaporisation   A process in which a liquid (or solid) changes into a vapour (gas).

viscosity   A description of how 'thick' a liquid is – for example engine oil is 'thicker' (more viscous) than petrol.

2.2

alkanes   A homologous series with general formula C_nH_{2n+2}. The simplest is methane, CH_4.

alkanoic acids   A homologous series in which a $-CH_3$ group in an alkane has been replaced by a carboxyl group $-COOH$. The simplest is methanoic acid HCOOH.

alkanols   A homologous series in which a hydrogen atom in an alkane has been replaced by a hydroxyl group, $-OH$. The simplest is methanol, CH_3OH.

alkenes   A homologous series with general formula C_nH_{2n}. The simplest is ethene, C_2H_4.

carboxyl group   The group $-COOH$ found in carboxylic acids such as the alkanoic acids.

carboxylic acid   One containing the carboxyl group $-COOH$.

cycloalkanes   A homologous series of ring molecules with general formula C_nH_{2n}. The simplest is cyclopropane, C_3H_6.

ester group   The group $-COO-$ found in ester molecules.

esters   Compounds formed in the reaction between an alkanol and an alkanoic acid – the simplest is methyl methanoate, $HCOOCH_3$.

functional group   A group of atoms with characteristic chemical activity – for example the bromo group, $-Br$, the hydroxyl group, $-OH$ and the carboxyl group, $-COOH$.

general formula   One from which the molecular formula for any member of a homologous series can be deduced.

homologous series   A group of chemically similar compounds which can be represented by a general formula. Physical properties change gradually through the series – for example the alkanes.

hydroxyl group   The group $-OH-$ for example in water and alkanols.

isomers   Compounds which have the same molecular formula but different structural formulae.

structural formula   One which shows the arrangement of atoms in a molecule or ion – a full structural formula shows all the bonds.

2.3

addition reaction   One in which two or more molecules join to produce a single larger molecule and nothing else.

catalytic hydration   An addition reaction in which water adds on to a molecule in the presence of a catalyst – for example ethene and water produce ethanol.

condensation reaction   One in which two or more molecules join to produce a single larger molecule along with water, or another small molecule.

cracking   The breaking up of large hydrocarbon molecules (usually alkanes) to produce a mixture of smaller molecules (usually alkanes and alkenes).

cracking (catalytic)   The use of a catalyst (in catalytic cracking) allows the process to take place at a lower temperature.

distillation   A process of separation or purification dependent on differences in boiling point. The changes of state involved are liquid to gas to liquid.

fermentation   The breakdown of glucose to produce ethanol and carbon dioxide catalysed by enzymes in yeast. It can also be applied to the similar breakdown of other organic molecules.

hydrolysis reaction   One in which a large molecule is broken down into two or more smaller molecules by reaction with water.

renewable energy resource   One which can be renewed and will therefore not run out in the foreseeable future – for example sugar cane.

reversible reaction   One which proceeds in both directions – for example, alkanol + alkanoic acid $\rightleftharpoons$ ester + water.

saturated hydrocarbon   One in which all carbon to carbon covalent bonds are single bonds, $C - C$.

unsaturated hydrocarbon   One in which there is at least one carbon to carbon double covalent bond, $C = C$.

2.4

addition polymerisation   A process in which many small monomer molecules join to form one large polymer molecule, and nothing else.

amide link   The group $- CONH -$ present in polyamides – it contains the same atoms as in the peptide link.

amine group   The group $- NH_2$ present in, for example, amines and amino acids – also called the amino group.

amine   A compound in which a hydrogen atom in an alkane, or other hydrocarbon, has been replaced by an amine group $- NH_2$.

biodegradable   Able to rot away by natural biological processes.

condensation polymerisation   A process in which many small monomer molecules join to form one large molecule, with (usually) water formed at the same time.

diamine   A compound in which two hydrogen atoms in an alkane, or other hydrocarbon, have been replaced by amine groups, $- NH_2$.

monomer   Relatively small molecules which can join together to produce a very large molecule (a polymer) by a process called polymerisation.

polyamides   Condensation polymers which can be made from dicarboxylic acids and diamines – they are collectively referred to as nylons.

polyester   A condensation polymer made from dicarboxylic acid and diol monomers.

polymer   A very large molecule formed by the joining of many small molecules called monomers.

polymerisation   A process in which a polymer molecule is formed.

thermoplastic   A plastic which softens on heating and can be reshaped – for example PVC and polythene.

thermosetting plastic   One which does not soften on heating – for example bakelite and formica.

toxic   Poisonous.

2.5

amino acid   Compounds containing the amine group $- NH_2$ and the carboxyl group $- COOH$. They undergo condensation polymerisation in the formation of proteins.

amylase   The enzyme which catalyses the hydrolysis of starch to glucose in the body.

Benedict's solution   A solution which turns from blue to orange-red when heated with many sugars – for example glucose, fructose and maltose (but not sucrose).

carbohydrates   A group of compounds with general formula $C_x(H_2O)_y$ – an important class of foods made by plants.

disaccharide   A carbohydrate of formula $C_{12}H_{22}O_{11}$: for example, sucrose, maltose.

fats and oils   An important group of high energy foods – they are esters of glycerol and fatty acids.

fatty acids   Saturated or unsaturated straight chain carboxylic (alkanoic) acids, usually with long chains of carbon atoms – for example stearic acid, $C_{17}H_{35}COOH$.

iodine solution   A solution which turns from reddish-brown to dark blue in the presence of starch.

monosaccharide   A carbohydrate of formula $C_6H_{12}O_6$: For example, glucose, fructose.

peptide link   The – CONH– group present in proteins – it contains the same atoms as in the amide link.

photosynthesis   The process by which plants make carbohydrates from carbon dioxide and water using light energy from the Sun in the presence of chlorophyll (as catalyst) – the process releases oxygen.

polysaccharide   A carbohydrate of formula $(C_6H_{10}O_5)n$, where n is a large number; examples are starch and cellulose.

proteins   An important class of foods made by plants – they all contain C, H, O and N and are condensation polymers made of many amino acids joined together.

respiration   The process occurring in plants and animals in which carbohydrates react with oxygen to release energy – carbon dioxide and water are also produced.

starch   A large carbohydrate molecule which is a natural condensation polymer made from glucose monomers.

sugars   Small carbohydrate molecules such as fructose and glucose, both $C_6H_{12}O_6$; and sucrose and maltose, both $C_{12}H_{22}O_{11}$.

(degree of) unsaturation   The extent to which a molecule has carbon–carbon double bonds.

ACIDS, BASES AND METALS

 Acids and bases

The pH scale

○ **pH** is a number that indicates the degree of acidity or alkalinity of a solution. The pH scale is a continuous range from below 0 to above 14:

−1　0　1　2　3　4　5　6　7　8　9　10　11　12　13　14　15
　　　　increasing acidity　　　　↑　　　　increasing alkalinity
　　　　　　　　　　　　　　　neutral

○ pH paper, pH indicator solution (**'universal' indicator solution**) or a pH meter can be used to find the pH of a solution.

○ **Acidic solutions** have a pH of less than 7; **alkaline solutions** have a pH of more than 7; pure water and neutral solutions have a pH equal to 7.

○ In pure water, and in all **neutral solutions**, there is a tiny but equal concentration of hydrogen and hydroxide ions, H^+(aq) and OH^-(aq) ions, respectively.

○ When **acids** dissolve in water they produce hydrogen ions, H^+(aq).

○ An acidic solution has a higher concentration of hydrogen ions than hydroxide ions.

○ When **alkalis** dissolve in water they produce hydroxide ions, OH^-(aq).

○ An alkaline solution has a higher concentration of hydroxide ions than hydrogen ions.

○ Diluting an acidic solution with water reduces the concentration of hydrogen ions, moving the pH *up* towards 7.

○ Diluting an alkaline solution with water reduces the concentration of hydroxide ions, moving the pH *down* towards 7.

○ In pure water and in aqueous solutions a **reversible reaction** takes place involving water molecules and hydrogen and hydroxide ions:

$$H_2O(l) \rightleftharpoons H^+(aq) + OH^-(aq)$$

At **equilibrium** in this reaction, there is a much greater concentration of water molecules than ions.

○ When a reversible reaction is at equilibrium, the concentrations of reactants and products remain *constant*, although not necessarily equal.

○ Non-metal oxides which dissolve in water produce *acidic* solutions – for example:

  – CO_2 gives carbonic acid, H_2CO_3

  – SO_3 gives sulphuric acid, H_2SO_4

  – NO_2 gives nitric acid, HNO_3.

○ Metal oxides and hydroxides which dissolve in water produce *alkaline* solutions. The oxides and hydroxides of Group 1 metals are all very soluble. For those in Group 2, solubility increases down the group.

○ Soluble metal hydroxides react with water producing the corresponding metal hydroxide – for example

$$\text{sodium oxide} + \text{water} \rightarrow \text{sodium hydroxide}$$
$$Na_2O + H_2O \rightarrow 2NaOH$$

○ **Lime water** is a common laboratory alkali – its chemical name is calcium hydroxide solution.

○ Ammonia gas is very soluble in water, dissolving to give an alkaline solution that is widely used in laboratories, in agriculture and in the home.

○ Acids and alkalis are in common use both in the laboratory and in the home.

Common laboratory acids

Name	Simple formula	Ions in aqueous solution
Hydrochloric acid	HCl	$H^+(aq)$ and $Cl^-(aq)$
Nitric acid	HNO_3	$H^+(aq)$ and $NO_3^-(aq)$
Sulphuric acid	H_2SO_4	$H^+(aq)$ and $SO_4^{2-}(aq)$

Common household acids

Name	Present in
Ethanoic acid	Vinegar
Citric acid	Citrus fruits – for example lemons
Carbonic acid	Fizzy drinks, soda water
Phosphoric acid	Cola drinks

Common laboratory alkalis

Name	Simple formula	Ions in aqueous solution
Sodium hydroxide	NaOH	$Na^+(aq)$ and $OH^-(aq)$
Potassium hydroxide	KOH	$K^+(aq)$ and $OH^-(aq)$
Calcium hydroxide	$Ca(OH)_2$	$Ca^{2+}(aq)$ and $OH^-(aq)$

Common household alkalis

Name	Present in
Sodium hydroxide	Oven cleaner and bleach
Sodium carbonate	Washing soda and dishwasher powder
Sodium hydrogencarbonate	Bicarbonate of soda

Concentration and the mole

❍ It is generally more useful to give the concentration of a solution in moles per litre ($mol\ l^{-1}$) than in grams per litre ($g\ l^{-1}$). For example, a $2\ mol\ l^{-1}$ solution contains 2 moles of solute per litre of solution.

❍ The relationship between *concentration of solution (C)*, *number of moles of solute (n)* and *volume of solution (V)* is

$$\text{concentration of solution} = \frac{\text{number of moles of solute}}{\text{volume of solution}}$$

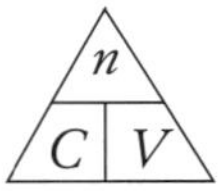

❍ The number of moles of solute, and the volume and concentration of solution can be calculated from the other two variables

$$n = C \times V; \quad V = \frac{n}{C}; \quad C = \frac{n}{V}$$

Example 1
Calculate the concentration of a citric acid solution, given that 1.5 moles of citric acid are dissolved in water and made up to a final volume of 3 litres.

$$\text{Concentration of a citric acid solution} = \frac{n}{V} = \frac{1.5}{3} = 0.5\ mol\ l^{-1}$$

Example 2

Calculate the concentration of a solution of sodium hydroxide, given that it contains 12 g of NaOH in 500 cm³ (0.5 litres) of solution.

$$\text{Number of moles of NaOH } (n) = \frac{\text{mass of NaOH in grams } (m)}{\text{formula mass of NaOH } (fm)}$$

$$= \frac{12}{40} = 0.3$$

$$\text{Concentration of NaOH solution, } C = \frac{n}{V} = \frac{0.3}{0.5} = 0.6 \text{ mol l}^{-1}$$

Example 3

Calculate the volume of solution produced if 53 g of sodium carbonate, Na_2CO_3, is used to make a solution with a concentration of 0.25 mol l¹.

$$\text{Number of moles of } Na_2CO_3, \ n = \frac{m}{fm} = \frac{53}{106} = 0.5$$

$$\text{Volume of } Na_2CO_3 \text{ solution, } V = \frac{n}{C} = \frac{0.5}{0.25} = 2 \text{ litres}$$

Preparation of a standard solution

○ A **standard solution** is one whose concentration is known accurately.

Aim

To prepare 100 cm³ of a 0.50 mol 1⁻¹ solution of sodium carbonate, Na_2CO_3.

Calculation

Number of moles of Na_2CO_3 required, $n = C \times V = 0.50 \times 0.1$
$$= 0.05$$

Mass of Na_2CO_3 required, $m = n \times fm = 0.05 \times 106 = 5.3$ g

Procedure

1 Weigh accurately 5.3 g of Na_2CO_3 in a small beaker.
2 Add a small volume of deionised (or distilled) water and stir to dissolve.
3 Transfer to a 100 cm³ standard flask along with washings from the beaker.
4 Make up to the mark carefully using more deionised (or distilled) water, using a dropper for the last few drops.
5 Invert several times to mix thoroughly.

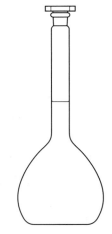

Figure 1 A standard flask

Strong and weak acids and alkalis

○ A **strong acid** is one which dissociates (breaks up) *completely* in aqueous solution to produce ions.
○ Hydrochloric, nitric and sulphuric acids are all strong acids:

$$HCl(aq) \rightarrow H^+(aq) + Cl^-(aq)$$
$$HNO_3(aq) \rightarrow H^+(aq) + NO_3^-(aq)$$
$$H_2SO_4(aq) \rightarrow 2H^+(aq) + SO_4^{2-}(aq)$$

○ A **weak acid** is one which dissociates only *partially* in aqueous solution to produce ions.

○ All carboxylic acids are weak, including ethanoic acid:

$$CH_3COOH(aq) \rightleftharpoons CH_3COO^-(aq) + H^+(aq)$$

○ *Equimolar* solutions (solutions of equal molar concentrations) of strong and weak acids differ in pH, conductivity and rate of reaction. This is because their hydrogen ion concentrations are different.

	1 mol l^{-1} HCl	1 mol l^{-1} CH$_3$COOH
pH	Well below 7	Below 7
Conductivity	High	Low
Reaction with Mg	Fast	Slow
Reaction with CaCO$_3$	Fast	Slow

○ A **strong alkali** is one which dissociates *completely* in aqueous solution to produce ions.

○ Soluble metal hydroxides are examples of strong alkalis. Unlike acids, which exist as polar molecules when pure, these alkalis are ionic solids. Dissolving simply frees the ions in the lattice. Sodium hydroxide is a strong alkali:

$$Na^+OH^-(s) + aq \rightarrow Na^+(aq) + OH^-(aq)$$

○ A **weak alkali** is one which dissociates only *partially* in aqueous solution to produce ions.

○ Ammonia solution is an example of a weak alkali:

$$NH_3(aq) + H_2O \rightleftharpoons NH_4^+(aq) + OH^-(aq)$$

○ Equimolar solutions of strong and weak alkalis differ in pH and conductivity because their hydroxide ion concentrations differ.

	1 mol l^{-1} NaOH	1 mol l^{-1} NH$_3$
pH	Well above 7	Above 7
Conductivity	High	Low

○ **Strength** and **concentration** are terms that are often used incorrectly. They do *not* have the same chemical meaning:

– strength is used to describe the extent of dissociation into ions

– concentration is used to describe the number of moles of solute in a certain volume of solution.

QUESTIONS

1 Look at this list of solutions of different types.

$H_2SO_4(aq)$ $NH_3(aq)$ $KNO_3(aq)$

$NaOH(aq)$ $MgCl_2(aq)$ $CH_3COOH(aq)$

a) Identify the strong acid.
b) Identify the weak acid.
c) Identify the strong alkali.
d) Identify the weak alkali.

2 Which of the following oxides dissolves in water to give a solution with a pH value greater than 7?
A CO_2
B CO
C SO_2
D CaO

3 Which of the following has the lowest pH value?
A 1 mol l^{-1} ammonia solution
B 1 mol l^{-1} hydrochloric acid
C 1 mol l^{-1} sodium hydroxide solution
D 1 mol l^{-1} ethanoic acid

4 Which of the following gases dissolves in water to give an alkaline solution?
A Nitrogen dioxide
B Ammonia
C Hydrogen chloride
D Nitrogen

5 The mass of sodium carbonate, Na_2CO_3, which is required to make 100 cm³ of a 0.1 mol l^{-1} solution is
A 106 g
B 10.6 g
C 1.06 g
D 0.106 g

6 A bottle is labelled '0.1 mol l^{-1} nitric acid'. This can best be described as:
A a concentrated solution of a weak acid
B a concentrated solution of a strong acid
C a dilute solution of a weak acid
D a dilute solution of a strong acid

7 Copy and complete the following using words and numbers of your own choice:
Neutral solutions contain equal concentrations of _____ and _____ ions and the pH of the solution is _____. In an acidic solution there are more _____ than _____ ions and the pH is _____ than _____. In an alkaline solution there are more _____ than _____ ions and the pH is _____ than _____.

8 a) If a metal oxide dissolves in water, what can be said about the pH of the solution produced?
b) If a non-metal oxide dissolves in water, what can be said about the pH of the solution produced?

9 Give the names and formulae of the negative ions present in aqueous solutions of the following:
a) hydrochloric acid
b) nitric acid
c) sulphuric acid

10 Write balanced formulae equations for the following reactions:
a) sulphur dioxide reacts *reversibly* with water to give sulphurous acid, H_2SO_3.
b) potassium oxide reacts with water to give potassium hydroxide.
c) magnesium oxide reacts with water to give magnesium hydroxide.

11 Calculate the number of moles of solute in each of the following solutions:
a) 3 litres of 0.5 mol l^{-1} potassium hydroxide
b) 0.4 litres of 2 mol l^{-1} nitric acid
c) 250 cm³ of 3 mol l^{-1} ammonia
d) 100 cm³ of 0.2 mol l^{-1} ethanoic acid

12 Calculate the mass in grams of solute present in each of the following solutions:
a) 2 litres of 0.5 mol l^{-1} $NaHCO_3$
b) 1.5 litres of 3 mol l^{-1} H_2SO_4
c) 600 cm³ of 0.8 mol l^{-1} NH_3
d) 15 cm³ of 0.02 mol l^{-1} $Ba(OH)_2$ (RAM of Ba = 137)

13 Calculate the concentration of each of the following solutions in moles per litre:
a) 4 moles of NaOH in 2 litres of solution
b) 303 g of KNO_3 in 6 litres of solution
c) 6.3 g of HNO_3 in 500 cm³ of solution
d) 30 g of H_2SO_4 in 200 cm³ of solution

14 Calculate the volume of solution that could be produced in each of the following cases:
a) 10 moles of HCl are used to prepare a 4 mol l^{-1} solution

b) 200 g of NaOH are used to prepare a 2.5 mol l^{-1} solution

c) 5.3 g of Na_2CO_3 are used to prepare a 0.2 mol l^{-1} solution

d) 20 g of $NaNO_3$ are used to prepare a 0.05 mol l^{-1} solution

15 a) Explain what is meant by the term *a strong acid*.

b) Ethanoic acid is described as a weak acid. Give a formula equation to show its dissociation in aqueous solution.

c) Explain why 1 mol l^{-1} hydrochloric acid, a strong acid, reacts faster with magnesium and calcium carbonate than 1 mol l^{-1} ethanoic acid, a weak acid.

16 a) Explain what is meant by the term *a weak alkali*.

b) Sodium hydroxide is described as a strong alkali. Give a formula equation to show its dissociation in aqueous solution.

c) Explain why 1 mol l^{-1} sodium hydroxide has a much higher conductivity than 1 mol l^{-1} ammonia solution.

17 a) In a solution of ethanoic acid, not all the molecules break up to form ions. What does this indicate about ethanoic acid?

b) In a solution of hydrochloric acid, all the hydrogen chloride molecules have broken up to form ions.

Two properties of solutions of ethanoic acid and hydrochloric acid were compared. The actual results for ethanoic acid are shown in the table.

	0.1 mol l^{-1} ethanoic acid	0.1 mol l^{-1} hydrochloric acid
pH	4	lower higher
Rate of reaction with magnesium	slow	slower faster

Which words in the right-hand column of the table show how the results for hydrochloric acid would compare with those for ethanoic acid?

(3.2) Salt preparation

Reactions of acids

○ **Neutralisation** is the reaction of acids with bases in which the pH of a solution moves towards 7.

○ **Bases** are compounds which react with acids to neutralise them – metal oxides, hydroxides, carbonates, hydrogencarbonates and ammonia are examples of bases.

○ Bases which dissolve in water form alkaline solutions and are called alkalis.

○ Neutralisation of a solution of an acid moves the pH up towards 7.

○ Neutralisation of a solution of an alkali moves the pH down towards 7.

Salts

○ **Salts** are ionic compounds formed in reactions between acids and bases, and between acids and metals. They contain a metal ion, or an ammonium ion, from the base or metal, and a negative ion from the acid.

Base/metal	Positive ion/ salt name	Acid	Negative ion/ salt name
Magnesium	Magnesium ...	Hydrochloric	... chloride
Sodium hydroxide	Sodium ...	Sulphuric	... sulphate
Calcium carbonate	Calcium ...	Nitric	... nitrate
Zinc oxide	Zinc ...	Ethanoic	... ethanoate

○ A salt is a compound in which the hydrogen ions of an acid have been replaced by metal ions or ammonium ions.

Alkali/acid reactions

○ The general word equation for an alkali/acid reaction is

$$\text{alkali} + \text{acid} \rightarrow \text{salt} + \text{water}$$

For example:

sodium hydroxide + hydrochloric acid → sodium chloride + water

$$NaOH(aq) + HCl(aq) \rightarrow NaCl(aq) + H_2O(l)$$

or $Na^+(aq) + OH^-(aq) + H^+(aq) + Cl^-(aq) \rightarrow Na^+(aq) + Cl^-(aq) + H_2O(l)$

Removing **spectator ions** (those which appear on both sides of the equation) gives:

$$OH^-(aq) + H^+(aq) \rightarrow H_2O(l)$$

❍ When acids and alkalis react, hydrogen ions and hydroxide ions join to form water.

Metal oxide/acid reactions

❍ The general word equation for a metal oxide/acid reaction is

$$\text{metal oxide} + \text{acid} \rightarrow \text{salt} + \text{water}$$

For example:

$$\text{zinc oxide} + \text{sulphuric acid} \rightarrow \text{zinc sulphate} + \text{water}$$
$$ZnO(s) + H_2SO_4(aq) \rightarrow ZnSO_4(aq) + H_2O(l)$$
$$\text{or } Zn^{2+}O^{2-}(s) + 2H^+(aq) + SO_4^{2-}(aq) \rightarrow Zn^{2+}(aq) + SO_4^{2-}(aq) + H_2O(l)$$

Removing spectator ions gives:

$$O^{2-}(s) + 2H^+(aq) \rightarrow H_2O(l)$$

❍ When acids and metal oxides react, hydrogen ions and oxide ions join to form water.

❍ It should be noted that the charge on a zinc ion *cannot* be worked out by reference to the Periodic Table since it is found among the transition metals. Fortunately, in all of the compounds in which it appears, it occurs as the zinc(II) ion, Zn^{2+}.

Metal carbonate/acid reactions

❍ The general word equation for a metal carbonate/acid reaction is

$$\text{metal carbonate} + \text{acid} \rightarrow \text{salt} + \text{water} + \text{carbon dioxide}$$

For example:

$$\text{sodium} + \text{nitric acid} \rightarrow \text{sodium nitrate} + \text{water} + \text{carbon}$$
$$\text{carbonate} \qquad\qquad\qquad\qquad\qquad\qquad \text{dioxide}$$
$$Na_2CO_3(aq) + 2HNO_3(aq) \rightarrow 2NaNO_3(aq) + H_2O(l) + CO_2(g)$$
$$\text{or } 2Na^+(aq) + CO_3^{2-}(aq) + 2H^+(aq) + 2NO_3^-(aq) \rightarrow 2Na^+(aq) + 2NO_3^-(aq) + H_2O(l) + CO_2(g)$$

Removing spectator ions gives:

$$CO_3^{2-}(aq) + 2H^+(aq) \rightarrow H_2O(l) + CO_2(g)$$

❍ When acids and carbonates react, hydrogen ions and carbonate ions react to form water and carbon dioxide.

Metal/acid reactions

❍ Not all metals react with acids but for those that do the general word equation for the reaction is

$$\text{metal} + \text{acid} \rightarrow \text{salt} + \text{hydrogen}$$

For example:

$$\text{magnesium} + \text{sulphuric acid} \rightarrow \text{magnesium sulphate} + \text{hydrogen}$$
$$Mg(s) + H_2SO_4(aq) \rightarrow MgSO_4(aq) + H_2(g)$$
$$\text{or } Mg(s) + 2H^+(aq) + SO_4^{2-}(aq) \rightarrow Mg^{2+}(aq) + SO_4^{2-}(aq) + H_2(g)$$

Removing spectator ions gives:

$$Mg(s) + 2H^+(aq) \rightarrow Mg^{2+}(aq) + H_2(g)$$

○ When acids and metals react, hydrogen ions are turned into hydrogen molecules.

○ The test for hydrogen is that it burns with a pop.

○ Only metals above hydrogen in the electrochemical series will displace hydrogen from an acid (see Section 3.3).

The damaging effects of acid rain

○ Sulphur dioxide, produced by the burning of fossil fuels (particularly coal) and nitrogen dioxide, produced by the sparking of air in car engines, dissolve in water to produce **acid rain**.

○ Sulphur dioxide dissolves to give sulphurous acid which is then converted (oxidised) to sulphuric acid. Nitrogen dioxide dissolves to give nitric acid.

○ The acidity in acid rain is mainly caused by the presence of sulphuric acid and, to a lesser extent, nitric acid.

○ Acid rain has damaging effects on buildings made from carbonate rocks – such as marble and limestone – structures made of iron and steel, soils and plant and animal life.

Everyday examples of neutralisation

○ Indigestion, caused by excess acid in the stomach, can be treated by using a suitable base to neutralise it – for example milk of magnesia (magnesium hydroxide) or bicarbonate of soda (sodium hydrogencarbonate).

○ Acidic soils can be treated with lime (calcium hydroxide) to improve conditions for plant growth.

○ Lochs which are too acidic can also be treated by adding lime.

Making fertilisers by neutralisation

○ Some soluble nitrogen-containing salts, including ammonium nitrate, ammonium sulphate and potassium nitrate, are made by neutralisation reactions. These salts are useful as fertilisers.

○ Ammonia reacts with nitric acid to give ammonium nitrate:

$$NH_3 + HNO_3 \rightarrow NH_4NO_3$$

○ Ammonia reacts with sulphuric acid to give ammonium sulphate:

$$2NH_3 + H_2SO_4 \rightarrow (NH_4)_2SO_4$$

Note: These are simplified equations – in each of them hydroxide ions from the alkali ammonia react with hydrogen ions from the acid to form water, in addition to the reaction shown.

○ Potassium hydroxide reacts with nitric acid to form potassium nitrate:

$$KOH + HNO_3 \rightarrow KNO_3 + H_2O$$

Making soluble salts by neutralisation of acids

Using alkalis

○ Pipette a known volume of the alkali into a conical flask and add a few drops of an indicator.

○ Add the acid from a burette until neutralisation is exact.

○ Note the volumes used and then mix these volumes *without* indicator.

○ Evaporate off some of the water.

○ Allow the salt to crystallise.

○ Filter off the crystals of salt.

Using insoluble bases or metals

○ In the preparation of a soluble salt, it is often easier to use an insoluble base or a metal – the insoluble base is usually a metal carbonate or metal oxide, such as copper(II) carbonate or copper(II) oxide.

○ Add the insoluble base or metal to the acid until no more reacts.

○ Filter off the excess base or metal.

○ Evaporate off some of the water.

○ Allow the salt to crystallise.

○ Filter off the crystals of salt.

Making insoluble salts by precipitation

○ **Precipitation** is the reaction of two solutions to form an insoluble product called a **precipitate**.

○ Insoluble salts can be formed by precipitation.

○ Select a soluble compound containing the positive ion of the required salt.

○ Select a soluble compound containing the negative ion of the required salt.

○ Mix solutions of the two compounds and filter off the insoluble salt – for example

$$\begin{array}{ccccccc} \text{zinc} & + & \text{sodium} & \rightarrow & \text{zinc} & + & \text{sodium} \\ \text{sulphate(aq)} & & \text{carbonate(aq)} & & \text{carbonate(s)} & & \text{sulphate(aq)} \end{array}$$

$$ZnSO_4(aq) + Na_2CO_3(aq) \rightarrow ZnCO_3(s) + Na_2SO_4(aq)$$

$$\text{or } Zn^{2+}(aq) + SO_4^{2-}(aq) + 2Na^+(aq) + CO_3^{2-}(aq) \rightarrow Zn^{2+}CO_3^{2}(s) + 2Na^+(aq) + SO_4^{2}(aq)$$

Removing spectator ions gives:

$$Zn^{2+}(aq) + CO_3^{2-}(aq) \rightarrow Zn^{2+}CO_3^{2-}(s)$$

Prescribed Practical Activity

Preparation of a salt

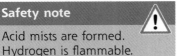

Safety note

Acid mists are formed.
Hydrogen is flammable.

- The aim of this experiment is to prepare a pure sample of a soluble salt – magnesium sulphate.
- Magnesium metal is added to a given volume of dilute sulphuric acid until there is no further evolution of hydrogen gas.

- The excess metal is filtered off and some of the water is evaporated from the salt solution obtained as the filtrate.
- The hot solution is set aside to crystallise.
- The same method can also be used with an insoluble base such as magnesium carbonate. In this case carbon dioxide is given off.
- Copper(II) carbonate can be reacted successfully to give copper(II) sulphate, but copper metal does *not* react with a dilute acid such as dilute sulphuric acid because it is too low in the electrochemical series.

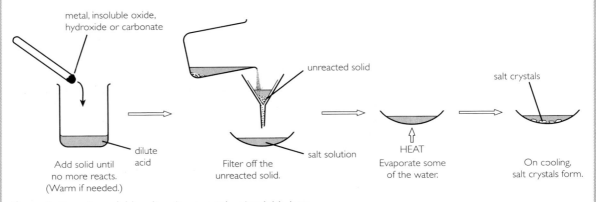

metal, insoluble oxide, hydroxide or carbonate

unreacted solid

salt crystals

dilute acid

Add solid until no more reacts. (Warm if needed.)

Filter off the unreacted solid.

salt solution

HEAT
Evaporate some of the water.

On cooling, salt crystals form.

Figure 1 Preparing soluble salts using a metal or insoluble base

Volumetric titrations

- ○ A **volumetric titration** is a procedure in which volumes of reacting solutions are measured. In acid/alkali titrations it is usual practice to use a pipette for measuring out the alkali and to add the acid from a burette.
- ○ The **end point** of the titration, when reaction is complete, is shown by a colour change in a suitable **indicator**.
- ○ Indicators are compounds whose colour depends on the pH of the solution in which they are present.
- ○ Normally, during a volumetric titration the concentration of one solution is known and the concentration of the other can be calculated based on knowing this, the volumes of the reacting solutions and a balanced equation for the reaction.
- ○ The relationship between concentration of solution (C), number of moles of solute (n) and volume of solution (V) can be used in calculations using volumetric titration results. This was introduced on page 87.

25
cm³

Figure 2 A pipette

Figure 3 A burette

Example 1

20 cm³ of sodium hydroxide solution was neutralised by 15 cm³ of 0.1 mol l⁻¹ sulphuric acid. Calculate the concentration of the sodium hydroxide solution.

The balanced equation for the reaction is:

$$H_2SO_4 + 2NaOH \rightarrow Na_2SO_4 + 2H_2O$$
$$\quad 1 \text{ mol} \qquad 2 \text{ mol}$$

Number of moles of acid reacting, $n_{H_2SO_4} = C \times V$

$$= 0.1 \times 0.015 = 0.0015$$

From the equation, 1 mol H_2SO_4 reacts with 2 mol NaOH.
So, 0.0015 mol H_2SO_4 reacts with 0.0030 mol NaOH.

So the concentration of NaOH, $C_{NaOH} = \dfrac{n_{NaOH}}{V}$

$$= \dfrac{0.003}{0.02} = 0.15 \text{ mol } l^{-1}$$

Alternative method

○ In titrations we usually know the volume of acid (V_A) and the volume of alkali or soluble base (V_B). We also know either the concentration of the acid (C_A) or the concentration of the alkali (C_B).

○ From the balanced equation for the reaction, we can find the number of moles of acid (a) and the number of moles of alkali (b) which react together.

○ Using the relationship $n = C \times V$, we can write the ratio of moles of acid to moles of alkali required for neutralisation as:

$$\frac{C_A \times V_A}{C_B \times V_B} = \frac{a}{b}$$

From this relationship we can calculate any of the six quantities if the other five are known.
Note: Because the relationship is a ratio, there is no need to convert the volumes into litres. Any units of volume may be used for V_A and V_B but they must be the same units in each case.

Example 2

Use the alternative method for the calculation given in the previous example.
The balanced equation for the reaction is:

$$H_2SO_4 + 2NaOH \rightarrow Na_2SO_4 + 2H_2O$$
$$\quad 1 \text{ mol} \qquad 2 \text{ mol}$$
$$\quad a = 1 \qquad b = 2$$

Rearranging $\dfrac{C_A \times V_A}{C_B \times V_B} = \dfrac{a}{b}$ gives $C_B = \dfrac{C_A \times V_A \times b}{V_B \times a}$

$$= \dfrac{0.1 \times 15 \times 2}{20 \times 1} = 0.15 \text{ mol } l^{-1}$$

QUESTIONS

1 Look at these six pairs of reactants.
  A $MgO(s) + HNO_3(aq)$
  B $AgNO_3(aq) + KCl(aq)$
  C $CuCO_3(s) + H_2SO_4(aq)$
  D $NaOH(aq) + HNO_3(aq$
  E $NH_3(aq) + H_2SO_4(aq$
  F $Zn(s) + HCl(aq)$

  Identify the pair(s) which react to produce:
  a) hydrogen gas
  b) carbon dioxide gas
  c) a precipitate
  d) a sulphate salt

2 Which of the following reactions is *not* an example of neutralisation?
  A $Mg(OH)_2(s) + 2HCl(aq) \rightarrow MgCl_2(aq) + 2H_2O(l)$
  B $CuO(s) + 2HNO_3(aq) \rightarrow Cu(NO_3)_2(aq) + H_2O(l)$
  C $BaCl_2(aq) + Na_2SO_4(aq) \rightarrow BaSO_4(s) + 2NaCl(aq)$
  D $Ca(OH)_2(aq) + H_2SO_4(aq) \rightarrow CaSO_4(aq) + 2H_2O(l)$

3 Which of the following compounds is a salt?
  A Potassium ethanoate
  B Calcium hydroxide
  C Aluminium oxide
  D Phosphorus trichloride

4 Which of the following pairs of solutions react to produce a precipitate?
  A Magnesium nitrate and sodium chloride
  B Ammonium carbonate and potassium nitrate
  C Barium chloride and calcium bromide
  D Lithium sulphate and barium nitrate

5 10 cm^3 of $2 \text{ mol } l^{-1}$ hydrochloric acid is exactly neutralised by
  A 10 cm^3 of $1 \text{ mol } l^{-1}$ sodium hydroxide solution
  B 5 cm^3 of $2 \text{ mol } l^{-1}$ sodium hydroxide solution

  C 5 cm^3 of $4 \text{ mol } l^{-1}$ sodium hydroxide solution
  D 10 cm^3 of $4 \text{ mol } l^{-1}$ sodium hydroxide solution

6 The balanced formula equation for the reaction between sodium hydroxide solution and dilute sulphuric acid is

  $$2NaOH + H_2SO_4 \rightarrow Na_2SO_4 + 2H_2O$$

  The volume of $0.1 \text{ mol } l^{-1}$ sodium hydroxide which is required to neutralise 20 cm^3 of $0.2 \text{ mol } l^{-1}$ sulphuric acid is
  A 20 cm^3
  B 40 cm^3
  C 60 cm^3
  D 80 cm^3

7 Copy and complete the following using words of your own choice:
  Neutralisation is the reaction of _____ with bases in which the _____ of the solution moves towards 7. Bases which dissolve in water are called _____. When a metal oxide or hydroxide reacts with an acid the products of the reaction are a _____ and _____. A salt is a compound in which the _____ ion of an acid has been replaced by a _____ ion or an ammonium ion.

8 Use words from the word bank to complete the word equations that follow.

  hydrogen   water   carbon dioxide   salt

  Each word may be used more than once if required.
  a) Metal hydroxide + acid → _____ + _____
  b) Metal oxide + acid → _____ + _____
  c) Carbonate + acid → _____ + _____ + _____
  d) Metal + acid → _____ + _____

9 Copy and complete the following word equations.
  a) Lithium hydroxide + hydrochloric acid → ...
  b) Potassium hydroxide + nitric acid → ...

c) Magnesium oxide + sulphuric acid → ...
d) Copper(II) oxide + hydrochloric acid → ...
e) Potassium carbonate + ethanoic acid → ...
f) Iron(II) carbonate + nitric acid → ...
g) Aluminium + hydrochloric acid → ...
h) Zinc + sulphuric acid → ...

10 For each of the following pairs of reactants:
   i) write a balanced formula equation in which no ions or state symbols are shown,
   ii) write a balanced ionic equation which shows state symbols with spectator ions 'cancelled'.
   a) Potassium hydroxide solution and dilute hydrochloric acid
   b) Sodium hydroxide solution and dilute nitric acid
   c) Magnesium oxide and dilute sulphuric acid
   d) Silver(I) oxide and dilute nitric acid
   e) Sodium carbonate solution and dilute hydrochloric acid
   f) Calcium carbonate and dilute hydrochloric acid
   g) Zinc and dilute sulphuric acid
   h) Aluminium and dilute hydrochloric acid

11 When solutions of zinc(II) sulphate and sodium carbonate are mixed, a white solid is formed.
   a) What name is given to this type of reaction?
   b) Name the insoluble white solid. (You may wish to refer to the data booklet.)
   c) Name the two ions from the reactants which remain in solution.
   d) What method should be used in order to separate the white solid from the solution?

12 Use the table of solubilities in the data booklet to help you complete the following equations using simple chemical formulae and state symbols.
   a) $MgCl_2(aq) + K_2CO_3(aq) \rightarrow MgCO_3(s) + \underline{\quad}$
   b) $Pb(NO_3)_2(aq) + 2NaI(aq) \rightarrow \underline{\quad} + \underline{\quad}$
   c) $Ba(NO_3)_2(aq) + Na_2SO_4(aq) \rightarrow \underline{\quad} + \underline{\quad}$
   d) $3LiCl(aq) + Na_3PO_4(aq) \rightarrow \underline{\quad} + \underline{\quad}$

13 Using ionic formulae and removing spectator ions, the equation in 12(a) can be rewritten as

$$Mg^{2+}(aq) + CO_3{}^{2-}(aq) \rightarrow Mg^{2+}CO_3{}^{2-}(s)$$

Write similar equations for questions 12(b), (c) and (d).

14 Given potassium hydroxide solution, dilute hydrochloric acid and any other chemicals or apparatus that you require, state how you would make crystals of potassium chloride.

15 Given insoluble copper(II) oxide, dilute sulphuric acid and any apparatus that you require, state how you would make crystals of copper(II) sulphate.

16 10 cm³ of 0.1 mol l⁻¹ sodium hydroxide neutralised 20 cm³ of hydrochloric acid – the balanced equation for the reaction being

$$NaOH(aq) + HCl(aq) \rightarrow NaCl(aq) + H_2O(l)$$

Calculate the concentration of the hydrochloric acid.

17 20 cm³ of 0.2 mol l⁻¹ potassium hydroxide neutralised 25 cm³ of nitric acid – the balanced equation for the reaction being

$$KOH(aq) + HNO_3(aq) \rightarrow KNO_3(aq) + H_2O(l)$$

Calculate the concentration of the nitric acid.

18 25 cm³ of 1 mol l⁻¹ sodium hydroxide neutralised 22.5 cm³ of sulphuric acid – the balanced equation for the reaction being

$$2NaOH(aq) + H_2SO_4(aq) \rightarrow Na_2SO_4(aq) + 2H_2O(l)$$

Calculate the concentration of the sulphuric acid.

19 10 cm³ of lithium hydroxide solution neutralised 16.7 cm³ of 0.1 mol l⁻¹ phosphoric acid – the balanced equation for the reaction being

$$3LiOH(aq) + H_3PO_4(aq) \rightarrow Li_3PO_4(aq) + 3H_2O(l)$$

Calculate the concentration of the lithium hydroxide solution.

20 Acid rain damages buildings made of carbonate rocks and structures made of iron and steel.
   a) Which two acids are the main causes of acid rain?
   b) Which two gases dissolve in rain water to produce these acids?
   c) i) Which of these gases is mainly produced by burning fossil fuels, particularly coal?
   ii) Which of these gases is produced by the sparking of air in petrol engines?

21 Fertilisers are used to improve crop yields. Give balanced formulae equations for the formation of the following fertilisers:

a) ammonium nitrate from ammonia and nitric acid

b) ammonium sulphate from ammonia and sulphuric acid

c) potassium nitrate from potassium hydroxide and nitric acid

d) ammonium phosphate from ammonia and phosphoric acid, H_3PO_4.

22 Vinegar is dilute ethanoic acid. The concentration of ethanoic acid in vinegar can be determined by neutralising a sample of vinegar with sodium hydroxide solution as shown in the diagram.

The volume of 0.5 mol l^{-1} sodium hydroxide solution required was 33.4 cm^3.
The equation for the reaction is

$$CH_3COOH(aq) + NaOH(aq) \rightarrow$$
$$CH_3COONa(aq) + H_2O(l)$$

Calculate the concentration, in mol l^{-1}, of the ethanoic acid in vinegar.

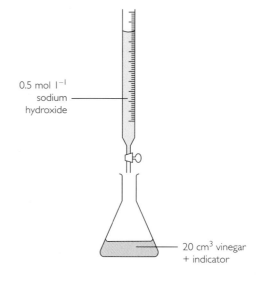

0.5 mol l^{-1} sodium hydroxide

20 cm^3 vinegar + indicator

(3.3) Metals

The electrochemical series

Batteries and cells

- In a chemical **cell**, chemical energy is converted into electrical energy.
- A **battery** consists of two or more cells joined together – although the words 'battery' and 'cell' tend to be used interchangeably.
- In a cell, one substance gives up electrons, another accepts them.
- An **electrolyte** is required to complete the circuit in a cell.
- Compared with mains electricity, batteries are safer and more portable but are more expensive and use more of finite resources such as zinc, lead and nickel.

Metals and the electrochemical series (ECS)

- Electricity can be produced by connecting two different metals with an electrolyte to form a simple cell, as shown in Figure 1.
- Different metals produce different voltages.
- The higher a metal is in the **electrochemical series**, the more readily it loses electrons.
- The further apart two metals are in the ECS, the greater is the cell voltage.
- In a cell, electrons flow from the metal higher in the ECS to the one lower down through the wires and meter.

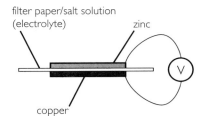

filter paper/salt solution
(electrolyte)   zinc

copper

Figure 1 A simple cell

Displacement and the ECS

- A **displacement reaction** is the formation of a metal from a solution containing its ions by reaction with a metal *higher* in the ECS – for example the displacement of copper from a solution of copper(II) ions by reaction with zinc:

$$Zn(s) + Cu^{2+}(aq) \rightarrow Zn^{2+}(aq) + Cu(s)$$

- The reaction between metals and acids can be used to establish the position of hydrogen in the ECS.
- Only those metals *above* hydrogen in the ECS can react with hydrogen ions and so displace the gas from an acid – for example

$$Fe(s) + 2H^{+}(aq) \rightarrow Fe^{2+}(aq) + H_2(g)$$

Oxidation, reduction and redox

- **Oxidation** is the loss of electrons during a reaction.
- **Reduction** is the gain of electrons during a reaction.
- A **redox reaction** is one in which both reduction and oxidation take place. Electrons are lost by one substance and gained by another – for example in displacement reactions such as

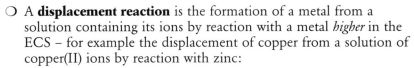

A useful mnemonic

Oxidation
Is
Loss (of electrons)

Reduction
Is
Gain (of electrons)

OIL RIG

101

$$Zn(s) + Cu^{2+}(aq) \rightarrow Zn^{2+}(aq) + Cu(s) \quad (redox)$$

The metal doing the displacing loses electrons and forms ions

$$Zn(s) \rightarrow Zn^{2+}(aq) + 2e^- \quad (oxidation)$$

The ions of the metal being displaced gain electrons and form metal atoms

$$2e^- + Cu^{2+}(aq) \rightarrow Cu(s) \quad (reduction)$$

The displacement of hydrogen from an acid solution is also an example of a redox reaction – for example

$$Fe(s) + 2H^+(aq) \rightarrow Fe^{2+}(aq) + H_2(g) \quad (redox)$$
$$Fe(s) \rightarrow Fe^{2+}(aq) + 2e^- \quad (oxidation)$$
$$2e^- + 2H^+(aq) \rightarrow H_2(g) \quad (reduction)$$

More about oxidation and reduction

○ A reaction in which a metal element forms a compound is considered to be an example of *oxidation.*

○ A reaction in which a compound forms a metal is considered to be an example of *reduction.*

More complex cells

○ Electricity can be produced in a cell by connecting two different metals in solutions of their ions. Electrons flow from the metal higher in the ECS to the one lower down through the wires and meter. In a cell containing magnesium and zinc, electrons flow from the magnesium to the zinc through the wires and meter.

Figure 2

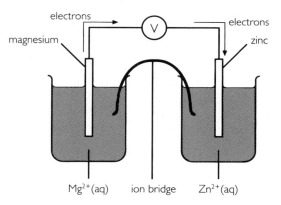

○ The metal higher in the ECS loses electrons and passes into solution as ions. The ions of the metal lower in the ECS gain electrons and form metal atoms. In the cell above, the following reactions take place at the metals:

$$Mg(s) \rightarrow Mg^{2+}(aq) + 2e^- \quad (oxidation)$$
$$2e^- + Zn^{2+}(aq) \rightarrow Zn(s) \quad (reduction)$$

○ All cells of this type must have an **ion bridge** – for example a piece of filter paper soaked in a salt solution. Its purpose is to complete the circuit by providing ions which are free to move into and out of the solutions.

○ Electricity can be produced in a cell when at least one of the half-cells does not involve metal atoms.

○ Carbon rods can be used to make contact with chemicals such as iodine molecules and sulphite ions, which can be used in half-cells. In a cell containing these species, electrons flow from the sulphite ions to the iodine through the wires and meter.

Figure 3

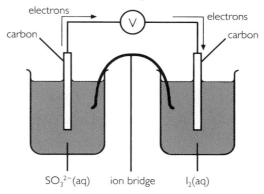

The reactions taking place in the half-cells are:

$$SO_3^{2-}(aq) + H_2O(l) \rightarrow SO_4^{2-}(aq) + 2H^+(aq) + 2e^- \quad \text{(oxidation)}$$
$$2e^- + I_2(aq) \rightarrow 2I^-(aq) \quad \text{(reduction)}$$

○ Species such as sulphite ions and iodine molecules can be included in the ECS, with sulphite ions above iodine molecules.

○ In a chemical cell, electrons flow in the *external* circuit (wires and meter) from the species higher in the ECS to the one lower down.

Prescribed Practical Activity

Factors which affect voltage

• The aim of this experiment is to investigate a factor which might affect the size of the voltage produced by a simple cell.

• One example of such a factor is the different pairs of metals used to produce different voltages.

• A cell like the one shown can be used.

• The zinc can be replaced by another metal, such as iron or magnesium, and the voltage noted again.

• Alternatively, zinc and copper can be retained and the electrolyte can be changed, the sodium chloride solution being replaced by hydrochloric acid or sodium hydroxide solution, for example.

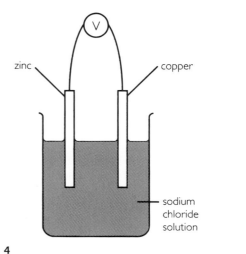

Figure 4

Redox equations from ion-electron equations

○ **Ion-electron equations** for the reduction and oxidation reactions taking place can be combined to give the redox equation for the overall reaction. In order to do this the number of electrons lost and gained must be the same.

○ If the ion-electron equations contain equal numbers of electrons, they are simply added to give the overall redox equation, with the electrons being omitted – for example the displacement reaction between zinc and copper(II) ions:

– write the ion-electron equations

$$\text{I} \quad Zn(s) \rightarrow Zn^{2+}(aq) + 2e^- \quad \text{(oxidation)}$$
$$\text{II} \quad 2e^- + Cu^{2+}(aq) \rightarrow Cu(s) \quad \text{(reduction)}$$

– add I and II (with the electrons cancelling out)

$$Zn(s) + Cu^{2+}(aq) \rightarrow Zn^{2+}(aq) + Cu(s) \quad \text{(redox)}$$

○ If the ion-electron equations do not contain equal numbers of electrons then the electrons are balanced by multiplying of one or both the ion-electron equations – for example the displacement reaction between copper and silver(I) ions:

– write the ion-electron equations

$$\text{I} \quad Cu(s) \rightarrow Cu^{2+}(aq) + 2e^- \quad \text{(oxidation)}$$
$$\text{II} \quad e^- + Ag^+(aq) \rightarrow Ag(s) \quad \text{(reduction)}$$

– multiply equation II by 2

$$2e^- + 2Ag^+(aq) \rightarrow 2Ag(s)$$

– add this to I (with the electrons cancelling out)

$$Cu(s) + 2Ag^+(aq) \rightarrow Cu^{2+}(aq) + 2Ag(s) \quad \text{(redox)}$$

○ The same process can be applied to more complex examples – for example the reaction between hydrogen peroxide and permanganate ions in acid solution:

– write the ion-electron equations (state symbols omitted for clarity)

$$\text{I} \quad H_2O_2 \rightarrow O_2 + 2H^+ + 2e^- \quad \text{(oxidation)}$$
$$\text{II} \quad 5e^- + MnO_4^- + 8H^+ \rightarrow Mn^{2+} + 4H_2O \quad \text{(reduction)}$$

– multiply equation I by 5 and equation II by 2 to give

$$5H_2O_2 \rightarrow 5O_2 + 10H^+ + 10e^-$$
$$10e^- + 2MnO_4^- + 16H^+ \rightarrow 2Mn^{2+} + 8H_2O$$

– add these equations (with the electrons cancelling out)

$$5H_2O_2 + 2MnO_4^- + 16H^+ \rightarrow 5O_2 + 10H^+ + 2Mn^{2+} + 8H_2O$$

– in this equation $10H^+$ ions may also be removed from both sides, giving the final overall equation

$$5H_2O_2 + 2MnO_4^- + 6H^+ \rightarrow 5O_2 + 2Mn^{2+} + 8H_2O \quad \text{(redox)}$$

Electrolysis as a redox process

○ During any electrolysis process, electron loss takes place at the positive electrode and electron gain at the negative electrode.

○ Oxidation takes place at the positive electrode where negative ions lose electrons.

○ Reduction takes place at the negative electrode where positive ions gain electrons.

○ During the electrolysis of copper(II) chloride solution, the reactions taking place at the electrodes are:

at the positive electrode negative chloride ions lose electrons and form chlorine molecules

$$2Cl^-(aq) \rightarrow Cl_2(g) + 2e^- \quad \text{(oxidation)}$$

at the negative electrode positive copper(II) ions gain electrons and form copper atoms

$$2e^- + Cu^{2+}(aq) \rightarrow Cu(s) \quad \text{(reduction)}$$

Reactions of metals

A reactivity series of metals

○ Metals can be placed in a **reactivity series** by observing their reactions with, for example, oxygen, water and dilute acids.

○ The most reactive metals are placed at the top of the reactivity series and the least reactive at the bottom.

○ Differences in reaction rates with a common reagent, such as dilute sulphuric acid, give an indication of the reactivity of metals.

○ Metals above mercury in the reactivity series react with oxygen when heated to produce the corresponding metal oxide – for example

$$\text{magnesium} + \text{oxygen} \rightarrow \text{metal oxide}$$
$$2Mg + O_2 \rightarrow 2MgO$$

○ Metals above aluminium in the reactivity series react with water to produce hydrogen gas and the corresponding metal hydroxide – for example

$$\text{sodium} + \text{water} \rightarrow \text{sodium hydroxide} + \text{hydrogen}$$
$$2Na + 2H_2O \rightarrow 2NaOH + H_2$$

○ Potassium, sodium and lithium are stored in oil because they react quickly with oxygen and water in the air.

○ All the metals above copper in the reactivity series react with dilute acid to produce a salt and hydrogen – for example

$$\text{zinc} + \text{hydrochloric acid} \rightarrow \text{zinc chloride} + \text{hydrogen}$$
$$Zn + 2HCl \rightarrow ZnCl_2 + H_2$$

○ Potassium, sodium and lithium are too reactive to risk in reactions with acids.

The ECS is a useful guide to the reactivity of metals	
Potassium	K
Sodium	Na
Lithium	Li
Calcium	Ca
Magnesium	Mg
Aluminium	Al
Zinc	Zn
Iron	Fe
Tin	Sn
Lead	Pb
Copper	Cu
Mercury	Hg
Silver	Ag
Gold	Au

○ Aluminium is slow to react because of its thin protective coating of aluminium oxide.

A summary of reactions of metals

Metal	Reaction with		
	Oxygen	Water	Dilute acid
Potassium Sodium Lithium Calcium Magnesium	Metal + oxygen ↓ metal oxide	Metal + water ↓ metal hydroxide + hydrogen	Metal + acid ↓ salt + hydrogen
Alumininum Zinc Iron Tin Lead		No reaction	
Copper			No reaction
Mercury Silver Gold	No reaction		

Prescribed Practical Activity

Reaction of metals with oxygen

- The aim of this experiment is to place three metals, such as magnesium, zinc and copper, in order of reactivity by observing their reaction when heated in oxygen.
- The apparatus is shown alongside.
- The metal is heated first, then the potassium permanganate and the metal are heated alternately, for a few seconds each, until a reaction involving the metal is observed.
- The higher a metal is in the reactivity series, the more vigorous is its reaction with oxygen.

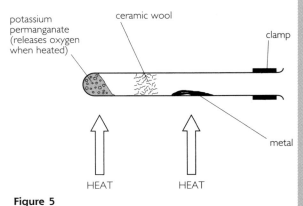

Figure 5

Extracting metals from their ores

○ Only unreactive metals like gold and silver occur uncombined in the Earth's crust.

○ Unreactive metals were the first to be discovered because their extraction did not involve a chemical reaction.

○ An **ore** is a naturally occurring compound of a metal from which more reactive metals must be extracted.

○ The extraction of a metal from its ore is an example of reduction.

○ Many ores are oxides, or can be easily converted to oxides from which the metal is extracted.

○ Reactive metals 'hold on' to oxygen more strongly than less reactive metals.

○ Oxides of metals below copper in the reactivity series can be decomposed to give the metal and oxygen gas by the use of heat alone – for example

$$\text{mercury(II) oxide} \rightarrow \text{mercury} + \text{oxygen}$$
$$2HgO \rightarrow 2Hg + O_2$$

○ Oxides of metals below aluminium in the reactivity series can be reduced to the metal by heating with carbon or carbon monoxide. These reactions take place because carbon bonds more strongly with oxygen than the metal does – for example

$$\text{zinc(II) oxide} + \text{carbon} \rightarrow \text{zinc} + \text{carbon dioxide}$$
$$2ZnO + C \rightarrow 2Zn + CO_2$$
$$\text{lead(II) oxide} + \text{carbon monoxide} \rightarrow \text{lead} + \text{carbon dioxide}$$
$$PbO + CO \rightarrow Pb + CO_2$$

○ Heating with carbon does not release the metal from the oxides of reactive metals (those above zinc in the reactivity series). This is because the metal bonds more strongly with oxygen than carbon does.

A summary of reactions of metal oxides

Metal	Effect of heating metal oxide	
	Alone	With carbon or carbon monoxide
Potassium Sodium Lithium Calcium Magnesium Aluminium	No reaction	No reaction
Zinc Iron Tin Lead Copper		Metal oxide + carbon or carbon monoxide ↓ metal + carbon dioxide
Mercury Silver Gold	Metal oxide ↓ metal + oxygen	

○ All the metals above zinc in the reactivity series are obtained by the electrolysis of molten compounds. Aluminium is obtained from its oxide, but the rest are obtained from their chlorides. The reaction taking place at the negative electrode during the electrolysis of molten aluminium oxide is

$$Al^{3+} + 3e^- \rightarrow Al$$

Extraction of iron in a blast furnace

○ Iron ore [iron(III) oxide], coke (carbon) and limestone (to remove impurities) are fed in at the top of a blast furnace.

○ A blast of air is fed in at the bottom of the furnace and the carbon undergoes incomplete combustion to give carbon monoxide:

$$\text{carbon} + \text{oxygen} \rightarrow \text{carbon monoxide}$$
$$2C + O_2 \rightarrow 2CO$$

○ The carbon monoxide reduces the iron(III) oxide to iron:

$$Fe_2O_3 + 3CO \rightarrow 2Fe + 3CO_2$$

○ The temperature in the blast furnace is so high that the iron produced is molten. It falls to the bottom of the furnace and is run off.

Corrosion

○ **Corrosion** is a chemical reaction which involves the surface of a metal changing from an element to a compound.

○ Corrosion of a metal is an example of oxidation.

○ Different metals corrode at different rates – reactive metals corrode at a faster rate than less reactive metals.

○ The term which is used to describe the corrosion of iron (or steel) is **rusting.**

○ **Rust** is the name of the compound formed when iron corrodes in air.

○ Both water and oxygen (from the air) are needed for iron to rust.

Rusting as a redox process

○ The first stage of rusting is the loss of electrons by iron to form iron(II) ions:

$$Fe(s) \rightarrow Fe^{2+}(aq) + 2e^- \quad \text{(oxidation)}$$

○ In solution, the iron(II) ions lose another electron to form iron(III) ions:

$$Fe^{2+}(aq) \rightarrow Fe^{3+}(aq) + e^- \quad \text{(oxidation)}$$

○ The electrons lost are gained by water and oxygen forming hydroxide ions:

$$4e^- + 2H_2O(l) + O_2(aq) \rightarrow 4OH^- (aq) \quad \text{(reduction)}$$

○ When $Fe^{3+}(aq)$ ions and $OH^-(aq)$ ions meets, rust forms. This can be thought of as either iron(III) hydroxide or a form of iron(III) oxide.

Detecting rusting

○ **Ferroxyl indicator** can be used to show when rusting is taking place.

○ $Fe^{2+}(aq)$ ions give a *blue* colour with ferroxyl indicator.

○ $OH^- (aq)$ ions give a *pink* colour with ferroxyl indicator.

Speeding up corrosion

○ Acid rain increases the rate of corrosion because first the acids present react with most metals – for example

$$Fe + H_2SO_4 \rightarrow FeSO_4 + H_2$$

and then the acids act as electrolytes in the chemical cells which are set up on the surface of metals during corrosion.

○ Salt (sodium chloride) spread on roads during winter speeds up the rusting of exposed iron and steel on cars. Like the acids in acid rain, all salts which dissolve in water act as electrolytes because they contain ions.

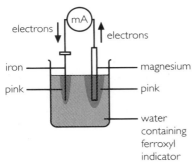

Figure 6 Magnesium/iron cell

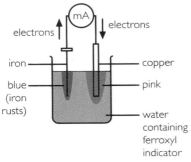

Figure 7 Copper/iron cell

Protection against corrosion

○ Coating a metal with paint, oil, grease, plastic or another metal provides *physical* protection because contact with air and water is prevented.

○ When a metal corrodes it loses electrons. This can be prevented by attaching the metal to the *negative* terminal of a battery or other d.c. supply. By doing this electrons flow *to* the metal, thus preventing corrosion.

○ A metal higher in the ECS can provide **sacrificial protection** to one lower down. When in contact, or joined by a wire, electrons can flow from the higher reactive metal to the lower one. The higher metal is *sacrificed* but the lower one is protected – for example magnesium is attached to iron pipes using wires, and zinc is attached directly to the iron hulls of ships.

○ In a magnesium/iron cell (Figure 6), electrons flow away from the magnesium (which is higher in the ECS) to the iron (which is lower). Magnesium is sacrificed and iron is protected.

  – at the magnesium: $Mg(s) \rightarrow Mg^{2+}(aq) + 2e^-$
  – at the iron: $4e^- + 2H_2O(l) + O_2(aq) \rightarrow 4OH^-(aq)$

○ In a copper/iron cell (Figure 7), electrons flow away from the iron (which is higher in the ECS) to the copper (which is lower). Iron is sacrificed and the copper is protected.

  – at the iron: $Fe(s) \rightarrow Fe^{2+}(aq) + 2e^-$
  – at the copper: $4e^- + 2H_2O(l) + O_2(aq) \rightarrow 4OH^-(aq)$

Galvanising iron

○ **Galvanising** involves putting a coating of zinc on to iron by dipping the iron object in molten zinc.

○ Galvanising provides *physical* protection for the iron by keeping out water and oxygen.

○ Galvanising, more importantly, provides *sacrificial* protection for iron because, even when the coating is broken, electrons can flow from the zinc (the metal higher in the ECS) to protect the iron.

○ A variety of iron and steel objects can be galvanised, from nails to car bodies.

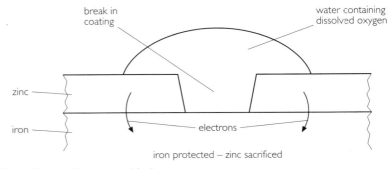

Figure 8 Galvanising iron with zinc

Tin-plated iron

○ Tin-plated iron is widely used for such items as food cans and biscuit tins. This is because tin is fairly unreactive and provides good *physical* protection, keeping water and oxygen away from the underlying iron.

○ When the tin coating is broken, the underlying iron rusts very rapidly. This is because iron is above tin in the ECS and electrons flow from the iron to the tin. The tin is protected, but the iron is sacrificed.

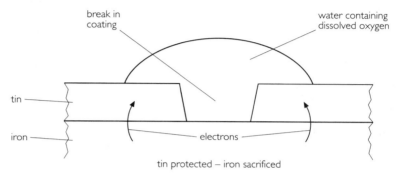

Figure 9 Tin-plated iron

Electroplating

○ **Electroplating** is a process in which a layer of metal – such as zinc, nickel or silver – is deposited on another metal by electrolysis.

○ The object to be electroplated is used as the negative electrode in a solution containing ions of the metal being deposited.

○ The positive electrode is usually made of the same metal as that being deposited at the negative electrode.

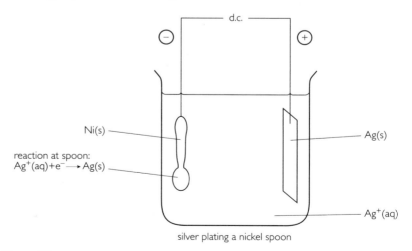

Figure 10

QUESTIONS

1 There are many types of chemical reactions. Examples of some of these are given in the list below.

A $2Cl^-(aq) \rightarrow Cl_2(g) + 2e^-$

B $Mg(s) + Zn^{2+}(aq) \rightarrow Mg^{2+}(aq) + Zn(s)$

C $Pb^{2+}(aq) + 2I^-(aq) \rightarrow Pb^{2+}(I^-)_2(s)$

D $5e^- + MnO_4^-(aq) + 8H^+(aq) \rightarrow Mn^{2+}(aq) + 4H_2O(l)$

E $SO_3^{2-}(aq) + H_2O(l) \rightarrow SO_4^{2-}(aq) + 2H^+(aq) + 2e^-$

F $Fe(s) + 2H^+(aq) \rightarrow Fe^{2+}(aq) + H_2(g)$

a) Identify the redox reaction(s).

b) Identify the oxidation reaction(s).

c) Identify the reduction reaction(s).

d) Identify the displacement reaction(s).

2 Simple cells were set up and attached to a voltmeter as shown below.

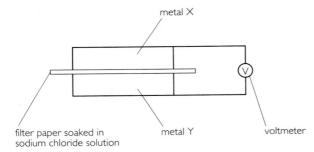

filter paper soaked in sodium chloride solution   metal Y   voltmeter

Which pair of metals would produce the highest reading on the voltmeter?

	Metal X	Metal Y
A	Zinc	Lead
B	Aluminium	Iron
C	Magnesium	Silver
D	Tin	Copper

3 A simple cell was set up as shown top right. Electrons flowed from metal X to metal Y through the wires and meter.

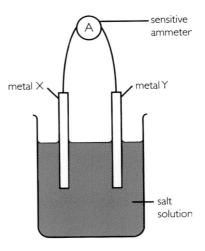

Which of the following correctly describes the metals used?

	Metal X	Metal Y
A	Zinc	Tin
B	Iron	Aluminium
C	Copper	Lead
D	Silver	Magnesium

4 Of three metals – X, Y and Z – only X must be extracted from its oxide ore using electrolysis. Y and Z can be extracted from their oxide ores by heating these with carbon. When Y and Z are in contact under corroding conditions the corrosion of Y is speeded up, but Z is protected from corrosion. The order of these metals in the electrochemical series, starting with the highest, is

A XYZ

B XZY

C ZYX

D YZX

5 Choose from this list to answer these questions:

A silver

B potassium

C lead

D zinc

a) Which of these metals can displace copper from copper(II) sulphate solution, but cannot displace iron from iron(II) sulphate solution?

b) Which metal cannot displace hydrogen from dilute hydrochloric acid?

c) Which metal is used as a coating for iron and steel, providing both physical and sacrificial protection from corrosion?

d) Which metal is most likely to be found uncombined in the Earth's crust?

6 Which of the following metal oxides is most easily reduced to the metal?
A Iron(III) oxide
B Tin(II) oxide
C Zinc(II) oxide
D Lead(II) oxide

7 In an investigation into corrosion the following experiment was set up.

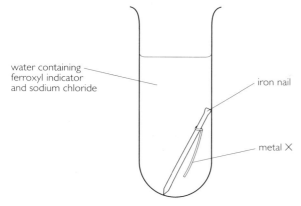

water containing ferroxyl indicator and sodium chloride

iron nail

metal X

After a few minutes a pink colour formed in the solution, but no blue colour was formed. Metal X could have been
A tin
B copper
C silver
D magnesium

8 a) In which of the following reactions will displacement take place?
A Zinc + magnesium nitrate solution
B Magnesium + iron(II) sulphate solution
C Lead + zinc sulphate solution
D Mercury + lead(II) nitrate solution

b) Where displacement takes place in part (a):
i) write an equation for the reaction taking place, omitting spectator ions.
ii) write ion-electron equations for the oxidation and reduction reactions involved.

9 When copper metal is placed in colourless silver(I) nitrate solution, the solution turns blue due to the formation of copper(II) ions and crystals of silver form on the copper.

a) Write a word equation for the reaction taking place.

b) Name the spectator ion in the reaction.

c) Give the relevant ion-electron equations associated with the reaction, labelling them 'oxidation' and 'reduction' as appropriate.

10 State the meaning of each of the following in terms of events involving electrons:
a) oxidation reaction
b) reduction reaction
c) redox reaction

11 a) Which metals, relative to the position of hydrogen in the electrochemical series, can displace hydrogen from an aqueous solution of an acid?

b) Write a balanced equation, complete with state symbols, for the reaction between zinc metal and hydrogen ions in an acid solution. State what type of reaction takes place.

c) Write ion-electron equations for the oxidation and reduction reactions involved.

12 During the electrolysis of molten zinc(II) chloride, zinc metal is formed at the negative electrode and chlorine gas at the positive electrode.
a) Write an ionic formula for zinc(II) chloride.
b) Give ion-electron equations for the reactions which take place at **(i)** the negative electrode and **(ii)** the positive electrode, adding 'oxidation' or 'reduction' as appropriate after each equation.

13 In the chemical cell shown, the reddish-brown colour caused by iodine molecules gradually fades.

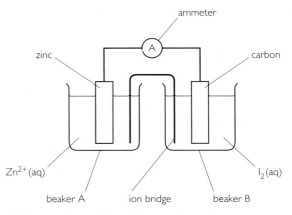

ammeter

zinc

carbon

Zn^{2+} (aq)

I_2 (aq)

beaker A

ion bridge

beaker B

The reaction taking place at the carbon electrode in beaker B is

$$2e^- + I_2(aq) \rightarrow 2I^-(aq)$$

a) What type of reaction is taking place at the carbon electrode in beaker B?

b) In which direction do the electrons flow through the wires and meter?

c) Give an ion-electron equation for the reaction which takes place at the zinc electrode in beaker A.

14 Many metals burn when heated in oxygen, combining with it to form an oxide.

a) How can this and other reactions be used to place a set of metals in a reactivity series?

b) What important factor, relating to the metals used, must be kept constant if the comparisons made are to be fair?

c) Write balanced formulae equations for the reaction of oxygen with **(i)** sodium, **(ii)** calcium and **(iii)** aluminium.

15 The alkali metals are so reactive that they are stored under oil.

a) Give the chemical symbols for all the alkali metals.

b) Which element present in air are the alkali metals likely to react with readily?

c) Which three alkali metals are less dense than water and therefore float on it when they are reacting?

d) Using the symbol M to represent an alkali metal, write a balanced formula equation for the reaction of M with water.

16 Use words from the word bank to copy and complete the paragraph given below:

zinc   silver   aluminium reactive

electrolysis unreactive carbon

Only _____ metals like gold and _____ are found uncombined in the Earth's crust. More _____ metals, such as iron and _____, are extracted from their oxides by heating with _____. The most reactive metals, including _____ are obtained by the _____ of one of their molten compounds.

17 Iron is extracted from iron(III) oxide ore in a blast furnace which is fed with ore, coke and limestone at the top, while a blast of air enters at the bottom. Most of the reduction to iron is brought about by reaction with carbon monoxide.

a) Explain how carbon monoxide is formed in a blast furnace.

b) Write a balanced equation for the reaction of iron(III) oxide with carbon monoxide.

c) Why is limestone fed into a blast furnace?

18 The metals of Groups 1 and 2 are extracted by electrolysis of their molten chlorides.

a) What is the meaning of the word 'electrolysis'?

b) Write ion-electron equations for the reactions taking place at the electrodes during the electrolysis of molten sodium chloride, adding 'oxidation' or 'reduction' after each equation as appropriate.

c) Why does solid sodium chloride not conduct electricity?

19 Corrosion is a chemical reaction which affects different metals to different extents.

a) What is the meaning of the word 'corrosion'?

b) Why does iron corrode at a faster rate than copper when the two metals are in the same environment?

c) Why is it incorrect to describe the corrosion of a metal such as aluminium as rusting?

d) Which two substances must normally be present for iron to corrode?

20 When an iron nail is placed in a test tube containing ferroxyl indicator solution, pink and blue areas of colour slowly develop in the liquid as the iron corrodes.

a) **i)** Which ion is responsible for the formation of the blue colour on reaction with ferroxyl indicator?
ii) Write an ion-electron equation for the formation of this ion.

b) **i)** Which ion is responsible for the formation of the pink colour on reaction with ferroxyl indicator?
ii) Write an ion-electron equation for the formation of this ion.

21 Dissolved compounds can speed up the corrosion of metals.

a) Give *two* reasons why the dissolved acids in acid rain speed up the corrosion of iron.

b) Explain why common salt, which is spread on icy roads during winter, speeds up corrosion of iron but sugar, used in a similar way, does not.

22 Steel car bodies are attached to the negative terminal of the car battery in an effort to slow down corrosion. In an investigation into this process the following experiment was set up.

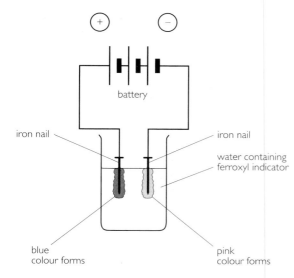

Explain *fully* why the results of this experiment support the decision of the motor car manufacturers to attach the negative terminal of the car battery to steel car bodies.

23 The steel used in the construction of the Forth road bridge was sprayed with molten zinc *after* construction had taken place.
   a) How are iron and steel objects normally coated with zinc?
   b) What name is given to the process of coating iron and steel with zinc?
   c) Explain why zinc can protect the steel of the bridge from corrosion even if the coating is broken and the steel is exposed.
   d) Why is this process of protection referred to as *sacrificial* protection?

24 Tin-plated steel, which is mainly iron, is widely used to make food containers.
   a) In what way does the tin provide *physical* protection of the underlying metal from corrosion?
   b) Why is corrosion of the iron very rapid, under corroding conditions, when the tin coating is broken?

25 Electroplating of one metal onto another can be used to provide a corrosion-resistant surface. Some motor car manufacturers prefer steel which has been electroplated with zinc because it provides a smooth surface for painting.
   a) Which electrode is the steel for electroplating purposes?
   b) Name a zinc compound which could be used as the electrolyte. You may wish to refer to the data booklet.
   c) Give an ion-electron equation for the reaction which would result in zinc being deposited on the chosen electrode.

26 A student set up the following cell.

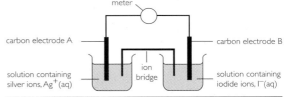

Electrode	Reaction taking place
A	$Ag^+(aq) + e^- \rightarrow Ag(s)$
B	$2I^-(aq) \rightarrow I_2(s) + 2e^-$

   a) Describe the path and direction of the electron flow in the cell.
   b) Combine the two ion-electron equations for the electrode reactions to produce a *balanced* redox equation.
   c) What is the purpose of the ion bridge?
   d) Describe the chemical test which could be used to show that iodine is formed at electrode **B**.

(3.4) Glossary

3.1

acid A compound which produces $H^+(aq)$ ions when it dissolves in water.

acidic solution One which contains a greater concentration of $H^+(aq)$ ions than pure water.

alkali A compound which produces $OH^-(aq)$ ions when it dissolves in water.

alkaline solution One which contains a greater concentration of $OH^?(aq)$ ions than pure water.

concentration The amount of solute in a given volume of solution. The usual units are moles per litre $(mol\ l^{-1})$.

equilibrium A situation when forward and reverse reactions in a reversible reaction take place at the same rate. The concentrations of reactants and products are constant but not necessarily equal.

lime water Calcium hydroxide solution – it is used to test for carbon dioxide, which turns it milky white.

neutral solution One in which the concentrations of $H^+(aq)$ and $OH^-(aq)$ are equal. The pH of a neutral solution is 7.

pH A number which indicates the degree of acidity or alkalinity of a solution. Acidic solutions pH $<$ 7; neutral solutions pH $=$ 7; alkaline solutions pH $>$ 7.

reversible reaction One which proceeds in both directions – for example, alkanol + alkanoic acid $\rightleftharpoons$ ester + water.

standard solution One whose concentration is known accurately.

strength The extent of dissociation into ions.

strong acid One which dissociates (breaks up) completely in aqueous solution to produce ions – for example hydrochloric, nitric and sulphuric acids.

strong alkali One which dissociates completely in aqueous solution to produce ions – for example sodium hydroxide and all other soluble metal hydroxides.

universal indicator solution A solution containing several indicators which can be used to give the approximate pH of a solution. It gives a range of colours depending on the pH of the solution.

weak acid One which dissociates (breaks up) only partially in aqueous solution to produce ions – for example ethanoic acid and all other carboxylic acids.

weak alkali One which dissociates only partially in aqueous solution to produce ions – for example ammonia solution.

3.2

acid rain Rain polluted by the presence of sulphuric and nitric acids.

base A substance which can accept hydrogen ions (from an acid) – for example ammonia and oxides, hydroxides and carbonates of metals.

end point The point of a volumetric titration when the reaction is complete which is shown by a colour change in a suitable indicator.

indicator A substance whose colour is dependent on pH.

neutralisation A reaction of acids with bases where the pH of a solution moves towards 7.

precipitate An insoluble solid which is formed on mixing certain solutions.

precipitation Reaction in which two solutions react to form an insoluble solid called a precipitate.

salt A compound in which the hydrogen ions of an acid have been replaced by metal ions or ammonium ions.

spectator ion An ion which is present in a reaction mixture but takes no part in the reaction.

volumetric titration An experiment in which volumes of reacting liquids are measured. In acid/alkali titrations it is usual practice to use a pipette for measuring out the alkali and to add the acid.

3.3

battery   A series of chemical cells joined together – the words 'cell' and 'battery' are often used interchangeably.

cell   In a chemical cell, chemical energy is converted into electrical energy. In an electrolytic cell, electrical energy is used to bring about chemical reactions at the electrodes.

corrosion   A chemical reaction in which the surface of a metal changes from an element to a compound.

decomposition   The breaking down of a compound into two or more substances (usually by heating) – for example $2HgO \rightarrow 2Hg + O_2$.

displacement reaction   Formation of a metal from a solution containing its ions by reaction with a metal higher in the ECS.

electrochemical series   A list of metals (and hydrogen) in order of their ability to lose electrons and form ions in solution – the ECS is similar to the reactivity series of metals.

electrolyte   A compound which conducts due to the movement of ions either when dissolved in water or melted.

electroplating   A process in which a layer of metal is deposited on another metal by electrolysis. The latter metal is used as the negative electrode in a solution containing ions of the metal being deposited.

ferroxyl indicator   Used to show the production of $Fe^{2+}(aq)$ ions (by the formation of a blue colour) and $OH^-(aq)$ ions (by the formation of a pink colour) during rusting of iron experiments.

galvanising   A process by which iron is coated with a protective layer of zinc.

ion bridge   Used to complete the circuit in a chemical cell by allowing a flow of ions through it.

ion-electron equation   One which shows either the loss of electrons (oxidation) or the gain of electrons (reduction).

ore   A naturally occurring compound of a metal from which metals can be extracted

oxidation   Reaction in which electrons are lost.

reactivity series   A list of metals in order of reactivity – for example with oxygen, water and dilute acids.

redox reaction   One in which reduction and oxidation take place. Electrons are lost by one reactant and gained by another.

reduction   Reaction in which electrons are gained.

rust   The name of the compound formed when iron corrodes. It can be looked upon as iron(III) hydroxide or a form of iron(III) oxide.

rusting   The corrosion of iron. Both water and oxygen are needed for iron to rust.

sacrificial protection   A method for protecting a metal from corrosion by attaching to it a metal which is higher in the ECS.

Whole-course questions

1 With the help of the data booklet, give the missing items in the table below.

Ion	Mass number	Atomic number	Number of protons	Number of neutrons	Number of electrons	Electron arrangement
$^{7}_{3}Li^{+}$	(a)	(b)	(c)	(d)	(e)	(f)
$^{23}_{12}Mg^{2+}$	(g)	(h)	(i)	(j)	(k)	(l)
(m)	27	(n)	13	(o)	(p)	2, 8

2 With the help of the data booklet, give the missing items in the table below.

Ion	Mass number	Atomic number	Number of protons	Number of neutrons	Number of electrons	Electron arrangement
$^{19}_{9}F^{-}$	(a)	(b)	(c)	(d)	(e)	(f)
$^{34}_{16}S^{2-}$	(g)	(h)	(i)	(j)	(k)	(l)
(m)	(n)	7	(o)	7	10	(p)

3 Barium metal can be obtained in two ways:
  I   electrolysis of molten barium chloride
  II heating aluminium with barium oxide (to give barium and aluminium oxide).

  a) i) Explain why barium chloride does not conduct electricity in the solid state.
  ii) The melting point of barium chloride is 963 °C. Give a reason for it being so high.

  iii) Give an ion-electron equation for the formation of barium during the electrolysis of molten barium chloride.
  b) i) Write a balanced equation for the reaction which takes place between aluminium and barium oxide when they are heated together.
  ii) Give a name for the type of reaction which takes place between aluminium and barium oxide.

4 Magnesium can be extracted from sea water.
One of the processes which has been used is
outlined below:

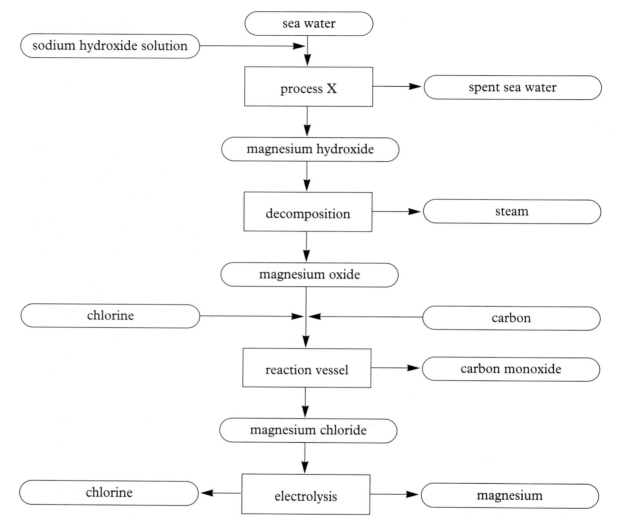

a) When sodium hydroxide solution is added
to sea water, a precipitate of magnesium
hydroxide is obtained.
   i) Write an ionic equation for this reaction
   showing state symbols, but omitting
   spectator ions.
   ii) Name the process X which separates
   magnesium hydroxide from the spent sea
   water.
b) Magnesium hydroxide is decomposed into
magnesium oxide and steam by heating.
Write a balanced equation for this reaction.
c) Magnesium oxide is heated with carbon
and chlorine to produce magnesium
chloride and carbon monoxide.

i) Write a balanced equation for this
reaction.
ii) Name another substance that magnesium
oxide could be reacted with in order to
produce magnesium chloride.
d) Electrolysis of molten magnesium chloride
produces magnesium and chlorine.
   i) Write ion-electron equations for the
   production of magnesium and chlorine,
   adding the words 'oxidation' and
   'reduction' as appropriate.
   ii) Explain why the company making the
   magnesium does not have to buy in
   chlorine.

5 Jean carried out an investigation into the rate of reaction of marble chips with dilute hydrochloric acid. She noted the loss in mass as carbon dioxide gas was given off, taking readings every minute. Her table of results is as follows.

Time /minutes	Mass of CO_2 produced/g
0	0
1	1.5
2	2.7
3	3.5
4	4.1
5	4.6
6	4.8
7	5.0
8	5.0

a) Draw a line graph to show these results.
b) Write a balanced equation for the reaction between marble chips (calcium carbonate) and dilute hydrochloric acid.
c) Calculate the average rate of reaction
   (i) over the first two minute period and
   (ii) over the second two minute period.
d) Give a reason to explain why the rate of reaction decreases.
e) In a separate experiment, using the same acid, the average rate of production of carbon dioxide was found to be 3.0 g min^{-1}. Give one change in the experiment which could have caused this result.

6 Ammonia is manufactured by the Haber Process in which nitrogen and hydrogen are passed over a hot iron catalyst under high pressure.

$$N_2(g) + 3H_2(g) \rightleftharpoons 2NH_3(g)$$

a) What do the arrows $\rightleftharpoons$ tell you about the reaction?
b) What *type* of catalysis takes place in the Haber Process?
c) The iron catalyst can be poisoned by certain impurities. Explain what happens during catalyst poisoning and why this results in a loss of catalyst efficiency.

7 Iron has been extracted from its ores for thousands of years, but aluminium production dates from much more recent times. Both are mainly extracted from their oxide ores.
a) Iron is obtained from iron(III) oxide in a blast furnace where coke (carbon) and limestone (calcium carbonate) are also present.
   i) A blast of air at the base of the furnace allows the coke to burn incompletely to form carbon monoxide. Write a balanced formula equation for this reaction.
   ii) The heat produced enables carbon monoxide to react with iron(III) oxide to give iron and carbon dioxide. Write a balanced formula equation for this reaction.
   iii) The limestone combines with sand (silicon dioxide), an impurity, forming calcium silicate and carbon dioxide. Write a balanced formula equation for this reaction given that the formula for the silicate ion is SiO_3^{2-}.
b) Aluminium must be extracted from aluminium oxide using electrolysis. A carbon-lined steel box is used for this purpose, with the lining acting as the negative electrode. The positive electrodes, which dip into the molten aluminium oxide in the box, are also made of carbon but these wear away quickly as oxygen is released.
   i) Why is solid aluminium oxide not electrolysed?
   ii) Write an ion-electron equation for the formation of aluminium at the negative electrode.
   iii) The operating temperature of the molten oxide is about 1000 °C. With the help of the data booklet, decide in what state the aluminium will be when it is formed.
   iv) Suggest a reason why the carbon positive electrodes wear away quickly.
c) Why is it more difficult to extract aluminium from aluminium oxide than to extract iron from iron(III) oxide?

8 Ethyne is an important hydrocarbon gas with molecular formula C_2H_2. It is better known by the older name of 'acetylene' and is widely used for cutting and welding steel as it gives a very hot flame when it burns in oxygen.
a) Draw a possible full structural formula for ethyne which obeys the normal valency rules.

121

b) Give a balanced formula equation for the complete combustion of ethyne.

c) Ethyne is the first member of a homologous series called the alkynes. The second member, propyne, has molecular formula C_3H_4, and the butynes – but-1-yne and but-2-yne – both have molecular formula C_4H_6.

i) Give the general formula for the alkynes.

ii) Give two further properties of a homologous series.

9 The alkanones are a subset of the set of organic compounds called carbonyl compounds. Some simple alkanones are shown below.

H   H
| |
H—C—C—C—H   propanone
| ‖ |
H   O   H

H   H   H
| | |
H—C—C—C—C—H   butanone
| ‖ | |
H   O   H   H

H   H   H   H
| | | |
H—C—C—C—C—C—H   pentan-2-one
| ‖ | | |
H   O   H   H   H

H   H   H   H
| | | |
H—C—C—C—C—C—H   pentan-3-one
| | ‖ | |
H   H   O   H   H

a) Give the molecular formula for pentan-3-one.

b) Give the general formula for the alkanones.

c) Name the following alkanone.

H   H   H   H   H
| | | | |
H—C—C—C—C—C—C—H
| ‖ | | | |
H   O   H   H   H   H

d) Hydrogen can add on to the C = O group in an alkanone, in the presence of a catalyst, to produce the corresponding alkanol. Name the alkanol which is produced when hydrogen undergoes an addition reaction with pentan-2-one.

10 Two hydrocarbon gases, X and Y, were found to have molecular formulae C_4H_8 and C_4H_6, respectively. Each was shaken with bromine solution with the following results:

I   X did not decolourise bromine solution

II Y decolourised bromine solution rapidly producing a colourless compound Z with molecular formula $C_4H_6Br_2$. This compound did not react further with bromine solution.

a) Give a possible name and full structural formula for hydrocarbon X.

b) Draw possible full structural formulae for compounds Y and Z.

11 Propan-2-ol can be made by the catalytic hydration of propene.

a) What is meant by the term 'catalytic hydration'?

b) Using full structural formulae, give the equation for this reaction.

c) Give the name and full structural formula of another alkanol which might be expected to be formed in this reaction in addition to the propan-2-ol.

12 The ester propyl ethanoate was prepared from an alkanol and an alkanoic acid.

a) Name the alkanol and alkanoic acid used in this reaction.

b) Name the catalyst used.

c) Give an equation for the reaction using full structural formulae for both reactants and products.

d) What type of reaction is involved in the formation of the ester?

13 Poly(ethenol) is an addition polymer which is soluble in water. A section of the polymer chain is shown below.

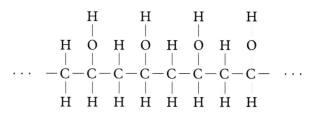

a) What is meant by the term 'addition polymer'?

b) Name the important functional group which is found in both the poly(ethenol) polymer and in water molecules.

c) How many repeating units are shown in the structure of the poly(ethenol) polymer above?

d) Draw the full structural formula for the repeating unit in poly(ethenol).

e) Draw the full structural formula for the monomer from which you would expect poly(ethenol) to be made, based on the repeating unit.

14 In Japan and Europe (excluding the UK) the most common polyamide fibre produced is nylon-6. Unlike nylon-6,6, which is made from two different monomers, nylon-6 is made from a single monomer. A part of the polymer is shown.

$$\cdots -N\begin{array}{c}H\\|\\\\|\\H\end{array}\left(\begin{array}{c}H\\|\\C\\|\\H\end{array}\right)_5\begin{array}{c}O\\||\\C\end{array}-N\begin{array}{c}H\\|\\\\|\\H\end{array}\left(\begin{array}{c}H\\|\\C\\|\\H\end{array}\right)_5\begin{array}{c}O\\||\\C\end{array}-N\begin{array}{c}H\\|\\\\|\\H\end{array}\left(\begin{array}{c}H\\|\\C\\|\\H\end{array}\right)_5\begin{array}{c}O\\||\\C\end{array}- \cdots$$

a) Draw the repeating unit in nylon-6.

b) Draw the full structural formula for the amide link.

c) Draw a full structural formula for the monomer from which nylon-6 is made.

d) What type of polymer is the polyamide nylon-6?

15 Several tests were carried out on carbohydrate X with the following results.

I   It dissolved readily in water.

II There was no colour change when a solution of X was heated in a water bath with Benedict's solution.

III There was no change in colour when a solution of X was tested with iodine solution.

IV When a solution of X was heated with a dilute acid it was discovered that one molecule of X reacted with one molecule of water to produce two different monosaccharide molecules, both of which gave a positive result when heated with Benedict's solution.

a) What can be deduced from test II alone?

b) What can be deduced from test III alone?

c) What type of reaction took place when a solution of X was heated with a dilute acid?

d) Describe the colour change when a substance gives a positive result on being tested with Benedict's solution.

e) Suggest names for the two monosaccharide molecules and give their molecular formulae.

f) Write a formula equation for the reaction of X with water.

g) Suggest a name for X.

16 The dipeptide shown below can be hydrolysed to produce two amino acids.

$$H-N\begin{array}{c}H\\|\\\end{array}C\begin{array}{c}H\\|\\\\|\\H-C-H\\|\\H\end{array}\begin{array}{c}O\\||\\C\end{array}-N\begin{array}{c}H\\|\\\end{array}C\begin{array}{c}H\\|\\\\|\\H-C-H\\|\\H-C-H\\|\\S\\|\\H-C-H\\|\\H\end{array}\begin{array}{c}O\\||\\C\end{array}-O-H$$

a) Draw full structural formulae for the amino acids formed on hydrolysis of the dipeptide.

b) The hydrolysis is carried out by refluxing an aqueous solution of the dipeptide with dilute hydrochloric acid – the hydrogen ions, $H^+(aq)$, catalyse the reaction. What type of catalysis is taking place?

c) Draw the full structural formula for the tripeptide which is formed when the amino acid shown below joins on to the *right*-hand side of the dipeptide.

$$H-N\begin{array}{c}H\\|\\\end{array}C\begin{array}{c}H\\|\\\\|\\H\end{array}\begin{array}{c}O\\||\\C\end{array}-O-H$$

d) Name the two functional groups which are present in all amino acids and give their full structural formulae.

17 Fats and oils are important components of our diet.

a) What is it that fats and oils can supply the body with about twice as much of compared to an equal mass of carbohydrate?

b) Hydrolysis of a fat produced the alkanol glycerol and several different fatty acids.
i) To which class of compounds do both fats and oils belong?
ii) Draw a structural formula for the alkanol glycerol.
iii) How many fatty acid molecules are combined with one molecule of glycerol in making a single fat molecule?

c) Oils can be converted into hardened fats. State what happens to oil molecules when this happens.

123

18 Some calcium was burned in air producing calcium oxide. This was then added to water producing 500 cm³ of a solution containing 2 g of calcium hydroxide.
   a) Give balanced formulae equations for these two reactions. There is no need to show ions or state symbols.
   b) Calculate the concentration of the calcium hydroxide solution in mol l⁻¹.
   c) Explain why calcium hydroxide is described as a strong alkali.

19 A student measured the pH of 0.1 mol l⁻¹ solutions of hydrochloric acid and ethanoic acid and obtained the following results.

Acid	Hydrochloric	Ethanoic
pH	1	3

   a) Which of the two solutions contains the higher concentration of hydrogen ions?
   b) Explain fully why the pH of the ethanoic acid is significantly higher than that of the hydrochloric acid.
   c) Write an equation for the dissociation of ethanoic acid in aqueous solution.

20 Plumbers sometimes use concentrated hydrochloric acid to remove rust marks from porcelain sinks and baths.
   a) i) Write a balanced formula equation for the reaction between rust, assumed to be iron(III) oxide, and hydrochloric acid. Use simple formulae and do not show any state symbols.
      ii) Rewrite the equation as an ionic equation, showing ions and state symbols.
   b) Before leaving the company that manufactures the hydrochloric acid, its concentration is checked by titration against a standard solution of sodium hydroxide in the following way:
   I the concentrated hydrochloric acid is diluted exactly 50 times
   II the diluted hydrochloric acid is titrated against standard 0.2 mol l⁻¹ sodium hydroxide solution.

   In one check it was found that 10 cm³ of 0.2 mol l⁻¹ sodium hydroxide solution required 9.2 cm³ of diluted hydrochloric acid for neutralisation. Calculate the concentration of the concentrated hydrochloric acid, given that the balanced equation for the neutralisation reaction is

$$NaOH(aq) + HCl(aq) \rightarrow NaCl(aq) + H_2O(l)$$

21 Sodium carbonate is an alkaline salt, a solution of which may be titrated against a strong acid such as sulphuric acid.
   a) Name the gas which is given off during the reaction.
   b) Name the salt which is formed in solution when sodium carbonate and sulphuric acid react.
   c) i) Write a balanced formula equation for the reaction of sodium carbonate solution with dilute sulphuric acid, using simple formulae only. There is no need to show state symbols.
      ii) Calculate the volume of 0.1 mol l⁻¹ sulphuric acid which would be needed to neutralise 20 cm³ of 0.12 mol l⁻¹ sodium carbonate solution.

22 The following cell was set up.

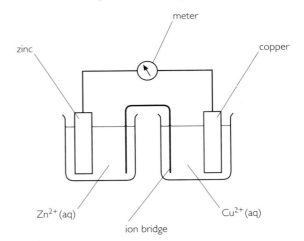

   a) In which direction will electrons flow through the wires and meter?
   b) Write ion-electron equations for the reactions taking place at the zinc and copper electrodes, adding the words 'reduction' or 'oxidation' as appropriate.
   c) Which electrode increases in mass?
   d) What would be the effect on the voltage that the cell could produce if the beaker containing copper and copper(II) ions was replaced by one containing a silver electrode dipping into a solution containing silver (I) ions?

23 Zeenat set up the following experiments to study corrosion involving iron nails. Each of the test tubes contained salt water and ferroxyl indicator.

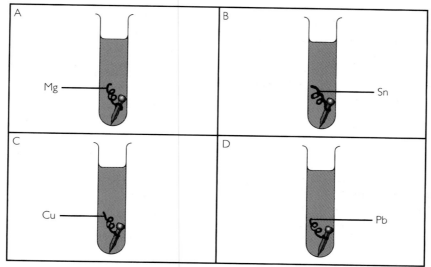

After a while, the following colours were observed.

Test tube	A	B	C	D
Colours	pink only	pink and blue	pink and blue	pink and blue

a) i) Which ion gives a pink colour with ferroxyl indicator?
   ii) Which ion gives a blue colour with ferroxyl indicator?
b) Explain fully why the fact that a piece of magnesium was attached to the iron nail in test tube A resulted in no blue colour being formed.
c) The results of the corrosion experiment in test tube B would suggest that it is not wise to attach tin to an iron object. Explain why tin-plated steel (steel is mainly iron), which is widely used for food cans, is
(i) satisfactory for this purpose while the tin coating is intact, but (ii) is not satisfactory when the tin coating is broken.
d) Oxygen and water molecules accept the electrons lost by iron when it corrodes. Write an ion-electron equation for this reduction process.

24 Look at these examples of different types of chemical reaction.
A $H_2SO_4 + CuO \rightarrow CuSO_4 + H_2O$
B $Mg + FeCl_2 \rightarrow MgCl_2 + Fe$
C $C_3H_6 + Br_2 \rightarrow C_3H_6Br_2$
D $C_4H_8 + 6O_2 \rightarrow 4CO_2 + 4H_2O$
E $C_{12}H_{26} \rightarrow C_2H_4 + C_{10}H_{22}$
F $CH_3COOH + CH_3OH \rightarrow CH_3COOCH_3 + H_2O$

a) Identify the following types of chemical reaction in the list:
   i) cracking
   ii) addition
   iii) condensation
   iv) neutralisation
   v) displacement
b) Name the sixth type of reaction which is given above.

25 Name the type of chemical reaction in each of the following examples. More than one descriptive term may be applicable in some cases.
a) $\dots CH_2 = CH_2 + CH_2 = CH_2 + \dots \rightarrow \dots - CH_2 - CH_2 - CH_2 - CH_2 - \dots$
b) $C_2H_4 + H_2O \rightarrow C_2H_5OH$
c) $C_{12}H_{22}O_{11} + H_2O \rightarrow 2C_6H_{12}O_6$
d) $C_6H_{12}O_6 \rightarrow 6C + 6H_2O$
e) $Zn(s) + 2H^+(aq) \rightarrow Zn^{2+}(aq) + H_2(g)$
f) $C_6H_{12}O_6 \rightarrow 2C_2H_5OH + 2CO_2$
g) $6e^- + Cr_2O_7^{2-} + 14H^+ \rightarrow 2Cr^{3+} + 7H_2O$
h) $\dots HOOCC_6H_4COOH + HOCH_2CH_2OH + HOOCC_6H_4COOH + HOCH_2CH_2OH + \dots \rightarrow \dots OCC_6H_4COOCH_2CH_2OOCC_6H_4COOCH_2CH_2O \dots + nH_2O$
i) $Na_2CO_3(aq) + Zn(NO_3)_2(aq) \rightarrow 2NaNO_3(aq) + ZnCO_3(s)$
j) $SO_3^{2-} + H_2O \rightarrow SO_4^{2-} + 2H^+ + 2e^-$
k) $(NH_4)_2Cr_2O_7 \rightarrow N_2 + Cr_2O_3 + 4H_2O$

4.2 Questions involving Prescribed Practical Activities

1 In an experiment to study the effect that varying the concentration of a reactant has on reaction rate, a solution containing sodium persulphate and starch was mixed with a solution containing potassium iodide and sodium thiosulphate. Persulphate ions and iodide ions react to produce iodine which, in turn, is converted back to iodide ions by reaction with thiosulphate ions.

a) What sharp change in colour of the solution marks the point when all of the thiosulphate ions have been used up?

b) In each experiment the time (t) from the mixing of the solutions until the sharp change in colour is noted. The results of one set of experiments were used to plot a graph of reaction rate ($1/t$) against the relative concentration of sodium persulphate solution which is shown in Figure 1.

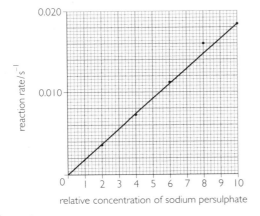

Figure 1

An error occurred when calculating the reaction rate for a relative concentration of 8 for the sodium persulphate.

i) Based on the graph, what should the reaction rate have been for this experiment?

ii) Calculate the time taken from the mixing of the solutions until the sharp colour change for this concentration of sodium persulphate.

c) Based on the graph, what conclusion can be reached concerning the reaction rate and the concentration of the sodium persulphate solution?

d) When carrying out the experiments, the total volume of the solutions reacting was kept constant. Name *two* other possible variables which must be kept constant throughout the experiments.

2 The reaction between sodium thiosulphate solution and dilute hydrochloric acid can be used to investigate the effect of a change in temperature on reaction rate. During the investigation the reaction mixture becomes more cloudy due to the formation of sulphur, and the time taken for a fixed mass of sulphur to form is noted.

a) State how the experiment is carried out in order to determine the time taken for a fixed mass of sulphur to form.

b) During a series of experiments, each carried out using a different temperature for the reaction mixture, the following results were obtained.

Temperature/°C	Time (t) for formation of a fixed mass of sulphur/s	Reaction rate $\left(\frac{1}{t}\right)$/s^{-1}
10	125	0.008
20	77	0.013
30	44	0.023
40	26	0.038
50	X	0.059

Draw a line graph of reaction rate against temperature.

c) Calculate the value for X in the table.

d) State four factors, concerning *only* the reacting solutions, which must be kept constant throughout the investigation.

3 When copper(II) chloride solution is electrolysed, chemical reactions take place at the electrodes.

a) Draw a labelled diagram of the apparatus used.

b) State what is observed at the negative electrode.

c) i) What is formed at the positive electrode?
ii) Describe the chemical test that you would carry out at the positive electrode in order to confirm your answer to part (c) (i). Give the results of your test.

d) Describe a hazard which is associated with the product formed at the positive electrode and a control measure which should be adopted in order to reduce it.

4 A hydrocarbon burns with a smoky yellow flame suggesting a high carbon content and that the hydrocarbon may be unsaturated.

a) What is meant by the term *unsaturated* in this context?

b) Describe how the test for unsaturation is carried out, and give the result of the test if the substance being tested is in fact unsaturated.

c) A hydrocarbon, X, with molecular formula C_4H_8 proved to be unsaturated. Draw two possible full structural formulae for this compound and give the name for each.

d) Another hydrocarbon, Y, with molecular formula C_4H_8 proved to be saturated. Draw a possible full structural formula for this compound and give its name.

e) Describe any hazard associated with the reagent used in the test for unsaturation and a control measure which can be used in order to reduce it.

5 The apparatus used in an experiment involving the cracking of the mixture of long chain alkanes called 'liquid paraffin' is shown in Figure 2. The object of the experiment is to show that the mixture of compounds produced contains unsaturated hydrocarbons.

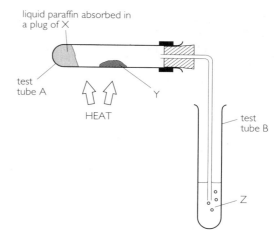

Figure 2

a) State what X, Y and Z are in the diagram.

b) State what is meant by the term 'cracking'.

c) A catalyst can be used to speed up a reaction, but that is not the reason why a catalyst is used in cracking reactions. State the correct reason.

d) What colour change is observed in the test tube containing Z when the products of cracking enter it?

e) Why must the delivery tube be removed from the liquid in test tube B *before* the heating of test tube A is stopped?

f) One of the alkanes present in liquid paraffin is octadecane, $C_{18}H_{38}$. If, on cracking, this breaks into two hydrocarbons, one of which is oct-1-ene which has molecular formula C_8H_{16}, give the name and molecular formula for the other compound formed.

g) Why is it not possible during the cracking of a large alkane molecule to obtain two smaller alkane molecules as the only products?

6 Figure 3 show two ways of hydrolysing starch.

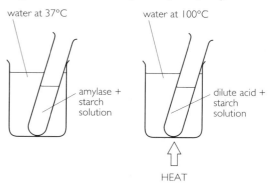

Figure 3

a) Describe the chemical test that would show the presence of starch. Give the result of the test.

b) In both cases, starch undergoes a hydrolysis reaction to form glucose. What is meant by the term 'hydrolysis reaction'?

c) Describe a chemical test which would confirm that glucose, or a similar sugar, had been produced. Describe the colour change during the test.

d) Name the enzyme that is used to hydrolyse starch.

e) Explain why it is important to also have a control test tube containing only starch solution in each beaker.

7 Crystals of magnesium sulphate can be prepared from the insoluble base magnesium carbonate and dilute sulphuric acid.

a) Name all the products of this reaction.

b) State how crystals of magnesium sulphate can be prepared using magnesium carbonate and 20 cm^3 of dilute sulphuric acid, dealing in turn with the following procedures:
- neutralisation (indicating three ways of knowing that neutralisation is complete)
- filtration
- evaporation
- crystallisation.

c) Explain what safety precaution would be needed if magnesium sulphate was made from magnesium and dilute sulphuric acid rather than magnesium carbonate and the same acid.

8 Various factors *might* affect the voltage of a simple cell, such as the one shown in Figure 4.

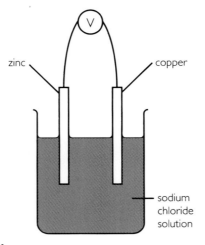

Figure 4

One factor which does affect the voltage is the use of different pairs of metals. In one set of experiments, the following readings were recorded.

Metal A	Metal B	Voltage/V
Iron	Copper	0.5
Aluminium	Copper	1.3
Zinc	Copper	0.9

a) How is the voltage produced related to the relative positions of the metals concerned in the electrochemical series?

b) Although correctly positioned by the voltage obtained, a higher result was obtained when aluminium was rubbed with sandpaper before being tested. What thin protective coating would be removed by the sandpaper, thus allowing better contact with the metal?

c) State three factors which should be kept constant when carrying out the experiments to see whether the use of different pairs of metals affects the voltage produced in a cell.

9 Some metals can be placed in order of reactivity by observing their reaction when heated in oxygen. The apparatus which is normally used is shown in Figure 5.

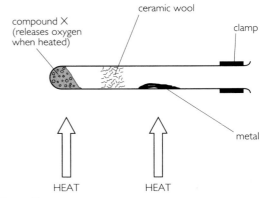

Figure 5

a) Name compound X, which releases oxygen when heated.

b) Why is it important to heat the metal *first* before heating compound X to release oxygen?

c) In one set of experiments the following results were obtained.

Metal	Observation
Copper	Changed from pink/brown to black, no glow
Magnesium	
Iron	No significant colour change, glowed brightly

State the observations that you would expect to be made for magnesium, making reference to the likely vigour of reaction and to the change in appearance.

d) In order that the comparisons are fair, state two factors which must be kept constant relating to the metals themselves.

Appendix

Types of chemical formulae – a summary

○ A **molecular formula** shows the number of atoms of the different elements which are present in one molecule of a substance – for example C_2H_4 for ethene.

○ An **empirical formula** shows the simplest whole-number ratio of atoms in a compound – for example SiO_2 for silicon dioxide.

○ An **ionic formula** shows the simplest whole-number ratio of ions in a compound – for example $Ca^{2+}(Cl)_2$ for calcium chloride.

○ A **full structural formula** shows all the bonds in a molecule or ion but does not necessarily show the true shape – for example for methane:

$$
\begin{array}{c}
H \\
| \\
H-C-H \\
| \\
H
\end{array}
$$

○ A **shortened structural formula** shows the sequence of groups of atoms in a molecule – for example $CH_3CH_2CH_2CH_3$ for butane.

○ A **perspective formula** shows the true shape of a molecule or ion – for example for methane (tetrahedral shape):

$$
\begin{array}{c}
H \\
| \\
C \\
H \diagup \quad \diagdown H \\
H
\end{array}
$$

○ What do you write when asked for 'the formula' or 'the simple formula' of a substance?

- For a **covalent molecular** substance we usually give the molecular formula – for example C_2H_6 for ethane. Sometimes, however, the empirical formula is used instead – for example P_2O_3 for phosphorus oxide when in fact the molecular formula is P_4O_6.

- For an **ionic** or **covalent network** compound we usually write the empirical formula, but show no charges in the ionic compound – for example NaCl for sodium chloride and SiO_2 for silicon dioxide.

- For **elements** a single symbol is usually written for all apart from those which exist as diatomic molecules (hydrogen, oxygen, nitrogen and the halogens) – for example C for carbon and Fe for iron, but H_2 for hydrogen and O_2 for oxygen. Only the noble gases exist as individual atoms not bonded to other atoms. In all other cases where single symbols are written as chemical formulae for elements, the symbol represents a sort of empirical formula.

QUESTIONS

1 The molecular formula for glucose is $C_6H_{12}O_6$.
   a) How many atoms are present in each glucose molecule?
   b) What is the empirical formula for glucose?

2 Silicon carbide is a covalent network compound in which the ratio of silicon atoms to carbon atoms is 1 : 1. What is the empirical formula of this compound?

3 Ammonia, which has molecular formula NH_3, has pyramid-shaped molecules. Draw a perspective formula for ammonia.

4 Citric acid is the chief acid in citrus fruits (6–7% in lemon juice). The structural formula of citric acid is

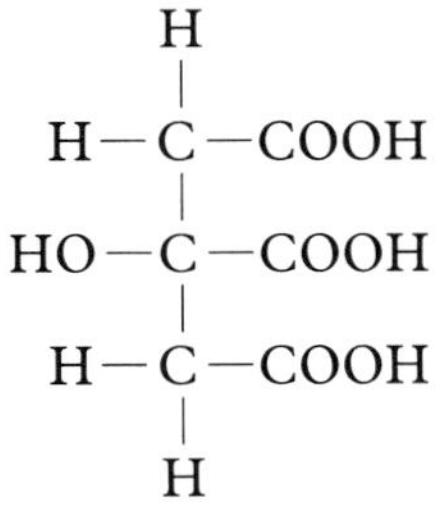

$$
\begin{array}{c}
\text{H} \\
| \\
\text{H}-\text{C}-\text{COOH} \\
| \\
\text{HO}-\text{C}-\text{COOH} \\
| \\
\text{H}-\text{C}-\text{COOH} \\
| \\
\text{H}
\end{array}
$$

What is the molecular formula of citric acid?

5 For each of the following give **(i)** the molecular formula, **(ii)** the empirical formula, **(iii)** the full structural formula and **(iv)** the shortened structural formula.
   a) ethane
   b) propene
   c) butane
   d) but-2-ene
   e) cyclohexane

6 Phosphorus molecules consist of four phosphorus atoms held together by covalent bonds. When phosphorus burns in a plentiful supply of oxygen, the main product consists of molecules in which four phosphorus atoms are bonded covalently with ten oxygen atoms. Use this information to write a balanced equation for the combustion of phosphorus using molecular formulae for the three substances involved.

7 The formulae of two common analgesics (pain-relief agents) are shown below:

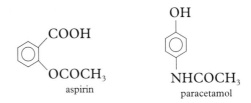

aspirin   paracetamol

In both formulae the groups

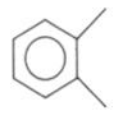

and    represent a group of six carbon atoms and four hydrogen atoms. Give molecular formulae for
   a) aspirin
   b) paracetamol.

8 Draw perspective formulae to show the true shapes of the following molecules. In each case give a word which describes the shape of the molecule concerned.
   a) Methane
   b) Water
   c) Tetrachloromethane, CCl_4
   d) Carbon dioxide

Section 5

ANSWERS

SECTION 1 BUILDING BLOCKS

1.1 **Substances**

1 B; **2** B; **3** A; **4** B; **5** A; **6** D

7 a) F, b) D, c) A and E, d) B

8 a) Iron, b) Oxygen, c) Krypton, d) Bromine

9 … Periodic Table … naturally … group … chemical properties

10 a) Carbon and sulphur.
  b) Hydrogen, beryllium, nitrogen, oxygen, fluorine, magnesium and chlorine.
  c) Neon.

11 a) Thallium and polonium are metals, tellurium is a non-metal.
  b) Astatine and francium do not occur naturally.
  c) Caesium (and francium).

12 a) …, lead, chlorine
    …, sodium, sulphur
    bromide, calcium, …
  b) …, zinc, carbon, oxygen
    copper, …, …, nitrogen, oxygen
    barium, …, …, sulphur, oxygen
    …, potassium, nitrogen, oxygen

13 There are many examples – vinegar, bleach, tea, coffee, lemonade etc.

14 a) D, b) C; **15** B

16 New substance formed (a white substance), change in appearance from silver to white, energy change/heat given out.

1.2 **Reaction rates**

1 C; **2** B

3 Increasing, decreasing, decreasing.

4 Since particles must collide for a reaction to take place:
  a) there are more particles in the same volume so the collision rate is increased
  b) the surface area for collision is increased and this increases the collision rate.

5 a) A catalyst is a substance which speeds up a reaction, but is not used up and can be recovered at the end of the reaction.
  b) A homogeneous catalyst is one which is in the same state as the reactants.
    A heterogeneous catalyst is one which is in a different state from the reactants.
  c) i) They are heterogeneous catalysts.
    ii) Carbon dioxide and nitrogen.
    iii) The lead compounds poison the catalyst.

6 a) Homogeneous.
  b) Enzymes catalyse reactions which take place in living cells.
  c) Many answers are possible – for example yoghurt and alcoholic drinks.

7 a) During the first 20 seconds the average
    rate $= \dfrac{12.5}{20} = 0.625 \text{ cm}^3 \text{ s}^{-1}$

    During the second 20 seconds the average
    rate $= \dfrac{10}{20} = 0.5 \text{ cm}^3 \text{ s}^{-1}$

During the third 20 seconds the average

rate $= \dfrac{7}{20} = 0.35$ cm^3 s^{-1}

b) Decrease in acid concentration and decrease in surface area of the marble chips as the reactants are used up.

c)

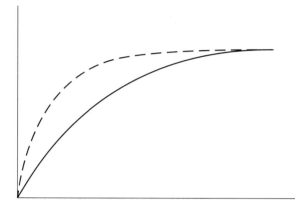

d) Rate $= \dfrac{\text{volume}}{\text{time}}$

time $= \dfrac{\text{volume}}{\text{rate}} = \dfrac{18 \text{ cm}^3}{2 \text{ cm}^3 \text{ s}^{-1}} = 9$ s

8 A cross is lightly drawn on a piece of paper. A known volume of sodium thiosulphate solution is put into a small beaker placed on the cross. A known volume of hydrochloric acid is added to the small beaker. The mixture is stirred and timing started. When the cross is no longer visible from above, the time for its disappearance is noted.

9 a) The concentration falls by $(0.010 - 0.004) = 0.006$ mol l^{-1}

b) Average rate

$= \dfrac{\text{concentration change}}{\text{time}}$

$= \dfrac{0.006}{400} = 0.000015$ mol l^{-1} s^{-1}

1.3 The structure of the atom

1 D; 2 a) C, b) B, c) D

3 D; 4 A; 5 C; 6 B

7 a) A and C, b) A and B, c) E and F, d) B and D

8 Copper must have two or more isotopes. The RAM is the average of the mass numbers of these isotopes, taking into account the proportions of each.

9 a) Isotopes are atoms of the same element which have different numbers of neutrons.
b) 6
c) $^{11}_{5}$B is present in greater amount because the relative atomic mass is closer to 11 than 10.

10 Nickel must have at least two isotopes. The relative atomic mass is an average of the mass numbers of these isotopes, taking into account the proportions of each.

11 a) Germanium, b) 32, c) 2, 8, 18, 4
d) Strontium, e) Sr, f) 2, 8, 18, 8, 2

12 a) 14, b) 6, c) 6, d) 8, e) 6, f) $^{21}_{10}$Ne, g) 10
h) 10, i) 11, j) $^{56}_{26}$Fe, k) 56, l) 26, m) 26

1.4 Bonding, structure and properties

1 a) H$_2$, b) KCl, c) H$_2$O and HF, d) C

2 B; 3 A; 4 C; 5 a) D, b) B; 6 B

7 a) If a d.c. supply was not used the sign of the electrodes would constantly change and the electrode products could not be identified.
b) Lead metal would form at the negative electrode and bubbles of chlorine gas at the positive electrode.
c) Negative electrode: Pb^{2+} + 2e$^-$ → Pb
Positive electrode: 2Cl$^-$ → Cl$_2$ + 2e$^-$

8 a) A covalent bond is the result of two positive nuclei being held together by their common attraction for a shared pair of electrons.
b) An ionic bond is the electrostatic force of attraction between oppositely charged ions.
c) A metallic bond is the electrostatic force of attraction between positive ions in a metallic lattice and the delocalised electrons within the structure.

9 a) 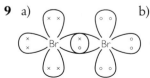 b)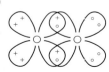

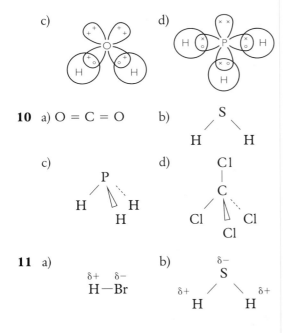

10 a) O = C = O   b)

$$S$$
$$H \quad H$$

c)

$$P$$
$$H \quad H$$
$$H$$

d)

$$Cl$$
$$C$$
$$Cl \quad Cl$$
$$Cl$$

11 a)

$$\delta+ \quad \delta-$$
$$H-Br$$

b)

$$\delta-$$
$$S$$
$$\delta+ \quad \delta+$$
$$H \quad H$$

12 a) Polar covalent bonds are formed when the attractions of the atoms for the bonded electrons are different.

b) The unusual properties of water include:
thin stream attracted to a charged rod
high boiling point
high freezing point.

13 Carbon dioxide is covalent molecular with only weak forces of attraction between its molecules. Silicon dioxide exists as a covalent network with strong covalent bonds holding its atoms together. At the melting point, it is covalent bonds in silicon dioxide which must be broken.

14 20 protons, 22 neutrons, 18 electrons arranged 2, 8, 8

1.5 Chemical symbolism

Chemical formulae

1 a) Na_2O, b) CaO, c) SiO_2, d) LiI,
e) $BaBr_2$, f) BN, g) PbO_2, h) $SnCl_4$,
i) $AlCl_3$, j) CBr_4, k) MgF_2, l) Mg_3N_2,
m) H_2S, n) NH_3, o) Ca_3P_2, p) KH

2 a) PbO, b) SO_3, c) ClO_2, d) N_2O,
e) PCl_3, f) CCl_4, g) C_3O_2, h) SF_6

3 Lead monoxide, sulphur trioxide, chlorine dioxide and sulphur hexafluoride. Tricarbon dioxide appears to disobey the rules, but does if the structure is $O = C = C = C = O$

It is also possible to draw some cyclic structures with molecular formula C_3O_2.

4 a) $LiNO_3$, b) $CaCO_3$, c) NH_4NO_3,
d) $CaCrO_4$, e) $BaSO_4$, f) $NaOH$,
g) Na_2CO_3, h) $CsOH$, i) NH_4I,
j) $BaCr_2O_7$, k) Li_3PO_4, l) $CaSO_3$,
m) $LiOH$, n) Na_2SO_4, o) NH_4NO_2,
p) Na_3PO_4, q) $MgSO_4$, r) $SrCO_3$,
s) $RbNO_3$, t) $BeSO_3$, u) KNO_3,
v) K_2SO_3, w) $NaHCO_3$, x) $KHSO_3$,
y) NH_4HSO_4

5 a) CuI, b) CuO, c) $FeCl_3$,
d) $FeBr_2$, e) Ag_2O, f) AuI_3,
g) $SnCl_4$, h) $PbBr_2$, i) V_2O_5,
j) CrF_3, k) $CoCl_2$, l) FeS,
m) $CuCl$, n) Ag_2S, o) ZnS,
p) PbO_2, q) SnH_4, r) TiN

6 a) $Ba(NO_3)_2$, b) $Ba(OH)_2$, c) $Ca_3(PO_4)_2$,
d) $Al(NO_3)_3$, e) $Mg(NO_2)_2$, f) $Ca(OH)_2$,
g) $Ra_3(PO_4)_2$, h) $(NH_4)_2SO_4$, i) $Sr(NO_3)_2$,
j) $Al_2(SO_4)_3$, k) $Al(OH)_3$, l) $Sr_3(PO_4)_2$,
m) $Ca(NO_3)_2$, n) $(NH_4)_2CO_3$, o) $Ca(HSO_4)_2$,
p) $Mg(HCO_3)_2$, q) $Ba(HSO_3)_2$

7 a) AgI, b) $PbCO_3$, c) $Zn(OH)_2$,
d) $Fe_2(SO_4)_3$, e) Ag_2CrO_4, f) Co_2O_3,
g) $Be(NO_3)_2$, h) $Fe(OH)_3$, i) $Ni(NO_3)_2$,
j) Ag_2SO_4, k) $Cu_3(PO_4)_2$, l) $Fe(NO_3)_3$,
m) Hg_2SO_4, n) $(NH_4)_3PO_4$, o) $Ca(MnO_4)_2$,
p) $CuCr_2O_7$

8 a) $(Na^+)_2O^{2-}$, b) $Ca^{2+}(Cl^-)_2$, c) K^+I^-,
d) $Ba^{2+}S^{2-}$, e) Li^+Br^-, f) $Mg^{2+}(F^-)_2$,
g) $Be^{2+}O^{2-}$, h) $Al^{3+}(F^-)_3$, i) $(K^+)_3P^{3-}$
j) $(Ca^{2+})_3(N^{3-})_2$, k) $(Mg^{2+})_3(P^{3-})_2$

9 a) $Na^+NO_3^-$, b) $K^+HCO_3^-$, c) $Ca^{2+}CO_3^{2-}$,
d) $NH_4^+Br^-$, e) Li^+OH^-, f) $Mg^{2+}CrO_4^{2-}$,
g) $Na^+MnO_4^-$, h) $Ca^{2+}Cr_2O_7^{2-}$, i) $K^+NO_3^-$,
j) $Na^+HSO_4^-$

10 a) $(Na^+)_2CO_3^{2-}$, b) $(K^+)_2SO_4^{2-}$,
c) $(NH_4^+)_2CO_3^{2-}$, d) $(NH_4^+)_2Cr_2O_7^{2-}$,
e) $(Li^+)_3PO_4^{3-}$, f) $Ca^{2+}(HCO_3^-)_2$,
g) $Ca^{2+}(OH^-)_2$, h) $Al^{3+}(OH^-)_3$,
i) $Ba^{2+}(HSO_3^-)_2$, j) $(Al^{3+})_2(SO_4^{2-})_3$,
k) $(NH_4^+)_3PO_4^{3-}$

11 a) Cu^+Cl^-, b) $Fe^{2+}SO_4^{2-}$,
c) $Cr^{3+}(F^-)_3$, d) $(Ag^+)_2SO_4^{2-}$
e) $Fe^{3+}(NO_3^-)_3$, f) $Cu^{2+}(OH^-)_2$,
g) $Ca^{2+}(HSO_3^-)_2$, h) $(Fe^{3+})_2(O^{2-})_3$,
i) $(Zn^{2+})_3(PO_4^{3-})_2$, j) $NH_4^+MnO_4^-$,
k) $Co^{3+}(F^-)_3$, l) $Ni^{2+}(NO_3^-)_2$

Chemical equations

1 a) $Na + O_2 \rightarrow Na_2O$
b) $Ca + O_2 \rightarrow CaO$
c) $P + O_2 \rightarrow P_2O_3$
d) $S + O_2 \rightarrow SO_2$
e) $C_3H_8 + O_2 \rightarrow CO_2 + H_2O$

2 a) $H_2 + F_2 \rightarrow HF$
b) $Hg + I_2 \rightarrow HgI_2$
c) $Fe + CuSO_4 \rightarrow Cu + FeSO_4$
d) $Mg + Zn(NO_3)_2 \rightarrow Zn + Mg(NO_3)_2$
e) $Ba + H_2O \rightarrow Ba(OH)_2 + H_2$

3 a) $Ca(OH)_2 + CO_2 \rightarrow CaCO_3 + H_2O$
b) $MgCO_3 + H_2O + CO_2 \rightarrow Mg(HCO_3)_2$
c) $BaO + H_2O \rightarrow Ba(OH)_2$
d) $Pb(NO_3)_2 + KI \rightarrow PbI_2 + KNO_3$
e) $NaOH + SO_2 \rightarrow Na_2SO_3 + H_2O$

Balancing chemical equations

1 $2Na + F_2 \rightarrow 2NaF$

2 $Mg + Cl_2 \rightarrow MgCl_2$

3 $N_2 + 3H_2 \rightarrow 2NH_3$

4 $4K + O_2 \rightarrow 2K_2O$

5 $H_2 + Br_2 \rightarrow 2HBr$

6 $Zn + H_2SO_4 \rightarrow ZnSO_4 + H_2$

7 $Ca + 2HCl \rightarrow CaCl_2 + H_2$

8 $Li_2O + H_2O \rightarrow 2LiOH$

9 $SnO_2 + 2H_2 \rightarrow Sn + 2H_2O$

10 $2KOH + H_2SO_3 \rightarrow K_2SO_3 + 2H_2O$

11 $Ca(OH)_2 + 2HCl \rightarrow CaCl_2 + 2H_2O$

12 $Ba(OH)_2 + H_2SO_4 \rightarrow BaSO_4 + 2H_2O$

13 $2Cu(NO_3)_2 \rightarrow 2CuO + 4NO_2 + O_2$

14 $MgCO_3 + 2HCl \rightarrow MgCl_2 + H_2O + CO_2$

15 $2C_3H_6 + 9O_2 \rightarrow 6CO_2 + 6H_2O$

16 $2C_6H_{14} + 19O_2 \rightarrow 12CO_2 + 14H_2O$

17 $2C_5H_{10} + 15O_2 \rightarrow 10CO_2 + 10H_2O$

18 $Fe_3O_4 + 4CO \rightarrow 3Fe + 4CO_2$

19 $PCl_3 + 3H_2O \rightarrow H_3PO_3 + 3HCl$

20 $NH_4NO_3 \rightarrow N_2O + 2H_2O$

1.6 The mole

Formula mass and the mole

1 a) 32, b) 64, c) 17, d) 58, e) 106,
f) 74, g) 149, h) 148, i) 342

2 Relationship used is $m = n \times fm$
a) 2 g, b) 44 g, c) 51 g, d) 100 g
e) 150 g, f) 43.5 g, g) 36.3 g, h) 375 g

3 Relationship used is $n = \dfrac{m}{fm}$

a) 3.5, b) 5, c) 1.25, d) 6.25, e) 0.0224,
f) 18.35, g) 8.065, h) 102

4 Mass in 16 tablets = $16 \times 0.5 = 8$ g
Number of moles of $CaCO_3 =$
$\dfrac{m}{fm} = \dfrac{8}{100} = 0.08$

5 Mass of butane = $10 \times 1000 = 10\,000$ g
Number of moles of butane =
$\dfrac{m}{fm} = \dfrac{10\,000}{58} = 172.4$

6 Mass of $(NH_4)_3PO_4 = 100 \times 1000 \times 1000 = 100\,000\,000$ g
Number of moles of ammonium phosphate =
$\dfrac{m}{fm} = \dfrac{10\,000\,000}{149} = 67\,114$

Calculations based on balanced equations

1 $C + O_2 \rightarrow CO_2$
1 mol $\leftrightarrow$ 1 mol
12 g $\leftrightarrow$ 44 g

480 g $\leftrightarrow \dfrac{44 \times 480}{12} = 1760$ g

2 $2Na + Cl_2 \rightarrow 2NaCl$
2 mol $\leftrightarrow$ 2 mol
46 g $\leftrightarrow$ 117 g

2 g $\leftrightarrow \dfrac{117 \times 2}{46} = 5.09$ g

3 $CaCO3 \rightarrow CaO + CO_2$
1 mol $\leftrightarrow$ 1 mol
100 g $\leftrightarrow$ 56 g
100 tonnes $\leftrightarrow$ 56 tonnes

5 tonnes $\leftrightarrow \dfrac{56 \times 5}{100} = 2.8$ tonnes

4 $CaCO_3 + 2HCl \rightarrow CaCl_2 + H_2O + CO_2$
1 mol $\leftrightarrow$ 1 mol
100 g $\leftrightarrow$ 44 g

$$20 \text{ g} \leftrightarrow \frac{44 \times 20}{100} = 8.8 \text{ g}$$

5 $Fe + 2HCl \rightarrow FeCl_2 + H_2$

1 mol $\qquad \leftrightarrow$ 1 mol

56 g $\qquad \leftrightarrow$ 2 g

56 mg $\qquad \leftrightarrow$ 2 mg

$$52 \text{ mg} \qquad \leftrightarrow \frac{2 \times 52}{56} = 1.86 \text{ mg}$$

6 $KOH + HNO_3 \rightarrow KNO_3 + H_2O$

1 mol $\qquad \leftrightarrow$ 1 mol

56 g $\qquad \leftrightarrow$ 101 g

56 tonnes $\qquad \leftrightarrow$ 101 tonnes

$$15 \text{ tonnes} \qquad \leftrightarrow \frac{101 \times 15}{56} = 27.05 \text{ tonnes}$$

7 $2NH_3 + H_2SO_4 \rightarrow (NH_4)_2SO_4$

2 mol $\qquad \leftrightarrow$ 1 mol

34 g $\qquad \leftrightarrow$ 132 g

34 kg $\qquad \leftrightarrow$ 132 kg

$$\frac{34 \times 500}{132} \text{ kg} \qquad \leftrightarrow 500 \text{ kg}$$

$$= 128.8 \text{ kg}$$

8 $Fe_2O_3 + 3CO \quad \rightarrow 2Fe + 3CO_2$

1 mol $\qquad \leftrightarrow$ 2 mol

160 g $\qquad \leftrightarrow$ 112 g

160 tonnes $\qquad \leftrightarrow$ 112 tonnes

$$\frac{160 \times 10}{112} \text{ tonnes} \leftrightarrow 10 \text{ tonnes}$$

$$= 14.3 \text{ tonnes}$$

9 $2H_2 + O_2 \quad \rightarrow 2H_2O$

2 mol $\qquad \leftrightarrow$ 2 mol

4 g $\qquad \leftrightarrow$ 36 g

$$\frac{4 \times 1000}{36} \text{ g} \leftrightarrow 1000 \text{ g}$$

$$= 111 \text{ g}$$

SECTION 2 CARBON COMPOUNDS

2.1 Fuels

1 a) i) Bitumen, ii) Fuel gas,
   b) Kerosene, c) Petrol

2 B; **3** D; **4** A

5 a) A fuel is a substance which is burned to produce energy.
   An exothermic reaction is one in which heat energy is released.
   b) Oxygen relights a glowing splint.

6 A hydrocarbon is a compound which contains hydrogen and carbon only.

7 a) Soot.
   b) Petrol engines reach a higher temperature (near the spark plug) resulting in more nitrogen dioxide being formed from nitrogen and oxygen in the air.

8 Fractional distillation ... fraction ... points ... increases ... attraction ... increase

2.2 Nomenclature and structural formulae

1 a) A and C, b) B, c) D, d) A, e) B

2 a) C_nH_{2n} b) i) $C_{22}H_{46}$ ii) $C_{13}H_{28}$

3 a)

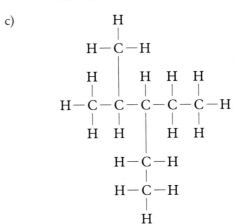

b)

c)

4 a) 3−methylpentane
b) 2,3,4−trimethylheptane
c) Cyclohexane

5 a)
```
     H   H   H   H   H   H
     |   |   |   |   |   |
 H — C — C — C = C — C — C — H
     |   |           |   |
     H   H           H   H
```

b)
```
     H   H   H   H   H   H   H   H
     |   |   |   |   |   |   |   |
 H — C — C = C — C — C — C — C — C — H
     |           |   |   |   |   |
     H           H   H   H   H   H
```

c)
```
  H        H   H   H   H   H   H
   \       |   |   |   |   |   |
    C = C — C — C — C — C — C — H
   /       |   |   |   |   |
  H        H   H   H   H   H
```

6 a) Propene
b) Pent-1-ene
c) Oct-4-ene

7 a) $CH_2 = CH_2$
b) $CH_3CH = CHCH_3$
c) $CH_3CH_2CH = CHCH_2CH_3$

8 a) A and E, b) E, c) D, d) F

9 a)
```
     H   H   H   H
     |   |   |   |
 H — C — C — C — C — H
     |   |   |   |
     H   O   H   H
         |
         H
```

b)
```
     H   H   H   H   H   H   H
     |   |   |   |   |   |   |
 H — C — C — C — C — C — C — C — O — H
     |   |   |   |   |   |   |
     H   H   H   H   H   H   H
```

c)
```
     H   H   H   H   H   H
     |   |   |   |   |   |
 H — C — C — C — C — C — C — H
     |   |   |   |   |   |
     H   H   O   H   H   H
             |
             H
```

10 a) Ethanol
b) Hexan-3-ol
c) Butan-1-ol

11 a)
```
     H       O
     |      //
 H — C — C
     |      \
     H       O — H
```

b)
```
     H   H   H       O
     |   |   |      //
 H — C — C — C — C
     |   |   |      \
     H   H   H       O — H
```

c)
```
     H   H   H   H   H       O
     |   |   |   |   |      //
 H — C — C — C — C — C — C
     |   |   |   |   |      \
     H   H   H   H   H       O — H
```

12 a) Methanoic acid
b) Propanoic acid
c) Pentanoic acid

13 a)
```
              O
             //
     H — C
             \        H
              O — C — H
                   |
                   H
```

b)
```
     H   H       O
     |   |      //
 H — C — C — C
     |   |      \         H   H
     H   H       O — C — C — H
                      |   |
                      H   H
```

c)
```
     H       O
     |      //
 H — C — C
     |      \         H   H   H
     H       O — C — C — C — H
                  |   |   |
                  H   H   H
```

14 a) Methyl ethanoate
b) Ethyl butanoate
c) Propyl methanoate

15 a) Ethyl ethanoate
b) Methyl pentanoate
c) Butyl propanoate

16 a) Methanol and propanoic acid
b) Ethanol and hexanoic acid
c) Propan-1-ol and ethanoic acid

17 D; **18** A

2.3 Reactions of carbon compounds

1 a) A and E, b) F, c) E, d) D, e) B, f) C

2 C; **3** D

4 a) Saturated hydrocarbons contain only carbon to carbon single covalent bonds, $C - C$.
Unsaturated hydrocarbons contain at least one carbon to carbon double covalent bond, $C = C$.
b) Unsaturated hydrocarbons decolourise bromine solution rapidly.

5 a) An alkane, b) i) Propane, ii) Ethene

6 a)

$$\begin{array}{ccccccc} & H & & H & & H & \\ & | & & | & & | & \\ H- & C & - & C & - & C & -H \\ & | & & | & & | & \\ & Br & & Br & & H & \end{array}$$

b)

$$\begin{array}{ccccccccc} & H & & H & & H & & H & & H \\ & | & & | & & | & & | & & | \\ H- & C & - & C & - & C & - & C & - & C & -H \\ & | & & | & & | & & | & & | \\ & H & & Br & & Br & & H & & H \end{array}$$

7 a) Cracking is a method of producing smaller, more useful molecules by heating large hydrocarbon molecules in the presence of a catalyst.
b) Cracking is important industrially because crude oil contains too many long chain hydrocarbons compared with the demand for them.
c) Aluminium oxide.
d) The carbon coating is burned off.
e) The catalyst allows cracking to take place at a lower temperature, saving energy and making the process more economical.

8 a) C_6H_{14}, b) H_2

9 Not enough hydrogen atoms are present.

10 a) An enzyme in yeast.
b) The yeast cells die if the ethanol concentration rises above about 12%.
c) By distillation.

11 a) Ethene.
b) Catalytic hydration.
c) By passing the ethanol vapour over hot aluminium oxide.

12 a) The reaction can go both ways.
b) Alkanols and alkanoic acids.

c) Mix equal volumes of an alkanol and an alkanoic acid in a test tube. Add a few drops of concentrated sulphuric acid and warm in a water bath. Pour the final mixture into a beaker of water – the ester floats on top.

13 a) In fruit-flavourings, perfumes and nail varnish remover – also as an industrial solvent.
b) Condensation.

14 a) A group of atoms with a characteristic chemical activity.
b)

2.4 Plastics and synthetic fibres

1 a) C, b) B and D, c) D and F, d) A and E

2 B; **3** B; **4** D; **5** B; **6** A

7 a) One which does not soften on heating and cannot be reshaped.
b) Electrical sockets and plugs can become hot, therefore thermoplastics would be unsuitable since they could soften or even melt.

8 a) Wool is warm and soft.
b) Cotton is cool and soft.
c) PVC is durable and does not rust.
d) Kevlar is strong and flexible.

9 a) A small molecule capable of joining with many others to form a polymer.
b) A very large molecule formed by the joining of many monomer molecules.
c) A process in which monomer molecules join to form one polymer molecule and nothing else.

d) A process in which many monomer molecules join to form one polymer molecule with water, or some other small molecule, formed at the same time.

10 a) Chloroethene

b)

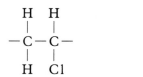

```
         H   H
         |   |
     — C — C —
         |   |
         H   Cl
```

c)

```
      H             H
       \           /
        C   =   C
       /           \
      H             Cl
```

11 a)

```
      H   CH₃  H   CH₃  H   CH₃
      |   |    |   |    |   |
··· — C — C —— C — C —— C — C —  ···
      |   |    |   |    |   |
      H   CH₃  H   CH₃  H   CH₃
```

b)

```
      H   CH₃
      |   |
    — C — C —
      |   |
      H   CH₃
```

12 a)

```
      H   O
      |   ||
    — N — C —
```

b)

```
       H  /H\  H   O  /H\  O
       |  |    |   ||  |   ||
    — N—⟨ C ⟩—N — C —⟨ C ⟩— C —
          |       |       |
          \H/₆        \H/₈
```

c)

```
   H      /H\      H        O     /H\      O
    \     |        /         \\    |        \\
     N —⟨ C ⟩— N          C —⟨ C ⟩— C
    /     |        \        /      |        \
   H      \H/₆      H   H—O  \H/₈   O—H
```

13 a) Condensation (it is also a polyester)

b)

```
         H  H    O         O     H  H    O         O
         |  |    ||        ||    |  |    ||        ||
··· — O — C — C — O — C — C₆H₄ — C — O — C — C — O — C — C₆H₄ — C —  ···
         |  |                    |  |
         H  H                    H  H
```

14 a)

```
   ⎧        H          H  O        O  H          H        ⎫
   ⎪        |          |  ||       ||  |          |        ⎪
   ⎨ —— N —(CH₂)₆— N — C —(CH₂)₄— C — N —(CH₂)₆— N —— ⎬
   ⎩                                                      ⎭
```

b) Both molecules have functional groups at each end of the molecule.

2.5 Natural products

1 a) C, b) B, c) A, d) F, e) D, f) E

2 B; **3** D

4 a) C, b) D, c) A, d) B

5 a) Only fructose gives an orange-red colour when heated with Benedict's solution.

   b) Only starch gives a dark blue colour with iodine solution.

6 a) Compounds which have the same molecular formula but different structural formulae.

   b) Glucose and fructose.

   c) Sucrose and maltose.

7 Carbon and hydrogen.

8 a) i) $(C_6H_{10}O_5)_n + nH_2O \rightarrow nC_6H_{12}O_6$

   ii) $C_{12}H_{22}O_{11} + H_2O \rightarrow C_6H_{12}O_6 + C_6H_{12}O_6$

   b) Enzymes.

9 a)

$$-\underset{\|}{\overset{O}{C}}-\underset{\|}{\overset{H}{N}}-$$

b)

$$H-\underset{\|}{\overset{H}{N}}-\underset{\underset{H}{|}}{\overset{H}{C}}-\underset{\|}{\overset{O}{C}}-O-H$$
$$\quad\quad\quad H-C-H$$
$$\quad\quad\quad\quad\quad H$$

$$H-\underset{\|}{\overset{H}{N}}-\underset{|}{\overset{H}{C}}-\overset{O}{\underset{\|}{C}}-O-H$$
with $H-C-H$, O, H chain

$$H-\underset{\|}{\overset{H}{N}}-\underset{|}{\overset{H}{C}}-\overset{O}{\underset{\|}{C}}-O-H$$
with $H-C-H$, H, $H-C-C-H$, H, $H-C-H$, H

10 a) i) Carbohydrates supply the body with energy.
ii) Proteins are the major structural materials of animal tissue.
iii) Fats supply the body with energy and are a more concentrated source of energy than carbohydrates.
b) Fats and carbohydrates contain C, H and O. Proteins contain C, H, O and N (and other elements).

11 a) The $C = C$ bond (they are unsaturated).
b) Oils will decolourise bromine solution.

12 a) A saturated or unsaturated straight chain carboxylic acid, usually with a long chain of carbon atoms.
b) 3 moles of fatty acids to 1 mole of glycerol.
c) $C_{17}H_{35}COOH$, $C_{13}H_{27}COOH$, $C_{17}H_{31}COOH$ and $CH_2OHCHOHCH_2OH$

SECTION 3 ACIDS, BASES AND METALS
3.1 Acids and bases

1 a) H_2SO_4, b) CH_3COOH, c) $NaOH$, d) NH_3

2 D; **3** B; **4** B; **5** C; **6** D

7 Hydrogen/hydroxide ... hydroxide/hydrogen ... 7 ... hydrogen ... hydroxide ... less ... 7 ... hydroxide ... hydrogen ... more ... 7

8 a) It will be greater than 7.
b) It will be less than 7.

9 a) Chloride, Cl^-
b) Nitrate, NO_3^-
c) Sulphate, SO_4^{2-}

10 a) $SO_2 + H_2O \rightleftharpoons H_2SO_3$
b) $K_2O + H_2O \rightarrow 2KOH$
c) $MgO + H_2O \rightarrow Mg(OH)_2$

11 Relationship used is $n = C \times V$
a) 1.5 b) 0.8, c) 0.75, d) 0.02

12 Relationship used is $n = C \times V$ and then $m = n \times fm$
a) $n = 0.5 \times 2 = 1$; $m = 1 \times 84 = 84$ g
b) 441 g, c) 8.16 g, d) 0.0513 g

13 Relationship used is $C = \dfrac{n}{v}$

a) 2 mol l^{-1}
b) $n = \dfrac{m}{fm} = \dfrac{303}{101} = 3$, $C = 0.5$ mol l^{-1}

c) 0.2 mol l^{-1}, d) 1.53 mol l^{-1}

14 Relationship used is $V = \dfrac{n}{C}$

a) 2.5 litre
b) $n = \dfrac{m}{fm} = \dfrac{200}{40} = 5$, $V = 2$ litre

c) 0.25 litre, d) 4.71 litre

15 a) One which dissociates completely in aqueous solution to produce ions.
b) $CH_3COOH(aq) \rightleftharpoons CH_3COO^-(aq) + H^+(aq)$
c) The hydrochloric acid has a higher concentration of hydrogen ions than the ethanoic acid.

16 a) One which dissociates only partially in aqueous solution to produce ions.
  b) $NH_3(aq) + H_2O(l) \rightleftharpoons NH_4^+(aq) + OH^-(aq)$
  c) The sodium hydroxide solution has a higher concentration of ions than the ammonia solution.

17 a) It indicates that it is a weak acid.
  b) For pH **higher**; for rate of reaction with magnesium **faster**.

3.2 Salt preparation

1 a) F, b) C, c) B, d) C and E

2 C; **3** A; **4** D; **5** C; **6** D

7 ... acids ... pH ... alkalis ... salt ... water ... hydrogen ... metal.

8 a) Salt + water
  b) Salt + water
  c) Salt + water + carbon dioxide
  d) Salt + hydrogen

9 a) ... lithium chloride + water
  b) ... potassium nitrate + water
  c) ... magnesium sulphate + water
  d) ... copper(II) chloride + water
  e) ... potassium ethanoate + water + carbon dioxide
  f) ... iron(II) nitrate + water + carbon dioxide
  g) ... aluminium chloride + hydrogen
  h) ... zinc sulphate + hydrogen

10 a) i) $KOH + HCl \rightarrow KCl + H_2O$
    ii) $\cancel{K}^+ (aq) + OH^- (aq) + \cancel{H}^+ (aq) + \cancel{Cl}^-$ (aq) $\rightarrow \cancel{K}^+ (aq) + \cancel{Cl}^- (aq) + H_2O(l)$
  b) i) $NaOH + HNO_3 \rightarrow NaNO_3 + H_2O$
    ii) $\cancel{Na}^+ (aq) + OH^- (aq) + H^+ (aq) + \cancel{NO}_3^-$ (aq) $\rightarrow \cancel{Na}^+ (aq) + \cancel{NO}_3^- (aq) + H_2O(l)$
  c) i) $MgO + H_2SO_4 \rightarrow MgSO_4 + H_2O$
    ii) $Mg^{2+} O^{2-}(s) + 2H^+(aq) + \cancel{SO}_4^{2-}$ (aq) $\rightarrow Mg^{2+} (aq) + \cancel{SO}_4^{2-} (aq) + H_2O(l)$
  d) i) $Ag_2O + 2HNO_3 \rightarrow 2AgNO_3 + H_2O$
    ii) $(Ag^+)_2 O^{2-}(s) + 2H^+(aq) + 2\cancel{NO}_3^-$ (aq) $\rightarrow 2Ag^+ (aq) + 2\cancel{NO}_3^- (aq) + H_2O(l)$
  e) i) $Na_2CO_3 + 2HCl \rightarrow 2NaCl + H_2O + CO_2$
    ii) $2\cancel{Na}^+ (aq) + CO_3^{2-}(aq) + 2H^+(aq) + 2\cancel{Cl}^- (aq) \rightarrow 2\cancel{Na}^+ (aq) + 2\cancel{Cl}^- (aq) + H_2O(l) + CO_2(g)$
  f) i) $CaCO_3 + 2HCl \rightarrow CaCl_2 + H_2O + CO_2$
    ii) $\cancel{Ca}^{2+} CO_3^{2-}(s) + 2H^+(aq) + 2\cancel{Cl}^- (aq) \rightarrow \cancel{Ca}^{2+} (aq) + 2\cancel{Cl}^- (aq) + H_2O(l) + CO_2(g)$
  g) i) $Zn + H_2SO_4 \rightarrow ZnSO_4 + H_2$

ii) $Zn(s) + 2H^+(aq) + \cancel{SO}_4^{2-} (aq) \rightarrow Zn^{2+}(aq) + \cancel{SO}_4^{2-} (aq) + H_2(g)$
  h) i) $2Al + 6HCl \rightarrow 2AlCl_3 + 3H_2$
    ii) $2Al(s) + 6H^+(aq) + 6\cancel{Cl}^- (aq) \rightarrow 2Al^{3+}(aq) + 6\cancel{Cl}^- (aq) + 3H_2(g)$

11 a) Precipitation
  b) Zinc(II) carbonate
  c) Sulphate ions and sodium ions
  d) Filtration

12 a) ... $2KCl(aq)$
  b) ... $PbI_2(s) + 2NaNO_3(aq)$
  c) ... $BaSO_4 (s) + 2NaNO_3(aq)$
  d) ... $Li_3PO_4(s) + 3NaCl(aq)$

13 b) $Pb^{2+}(aq) + 2I^-(aq) \rightarrow Pb^{2+}(I^-)_2(s)$
  c) $Ba^{2+}(aq) + SO_4^{2-}(aq) \rightarrow Ba^{2+}SO_4^{2-}(s)$
  d) $3Li^+(aq) + PO_4^{3-}(aq) \rightarrow (Li^+)_3PO_4^{3-}(s)$

14 Pipette a known volume of the potassium hydroxide solution into a conical flask and add a few drops of an indicator.
  Add the hydrochloric acid from a burette until neutralisation is exact.
  Note the volumes used and then mix these volumes without indicator present.
  Evaporate off some of the water. Allow to crystallise. Filter off the crystals.

15 Add copper(II) oxide to a given volume of dilute sulphuric acid until no more reacts.
  Filter off the excess, then evaporate off some of the water and allow to crystallise.
  Filter off the copper(II) sulphate crystals.

16 $n_{NaOH} = C \times V = 0.1 \times 0.01 = 0.001$
  From the equation 1 mol of NaOH reacts with 1 mol of HCl.
  Thus 0.001 mol of NaOH reacts with 0.001 mol of HCl.
  $C_{HCl} = \dfrac{n}{V} = \dfrac{0.001}{0.02} = 0.05$ mol l^{-1}

17 $n_{KOH} = C \times V = 0.2 \times 0.02 = 0.004$
  From the equation 1 mol of KOH reacts with 1 mol of HNO_3.
  Thus 0.004 mol of KOH reacts with 0.004 mol of HNO_3.
  $C_{HNO} = \dfrac{n}{V} = \dfrac{0.004}{0.025} = 0.16$ mol l^{-1}

18 $n_{NaOH} = C + V = 1 \times 0.025 = 0.025$
  From the equation 2 mol of NaOH react with 1 mol of H_2SO_4.
  Thus 0.025 mol of N$_a$OH react with 0.0125 mol of H_2SO_4.

$$C_{H2SO4} = \frac{n}{V} = \frac{0.0125}{0.0225} = 0.556 \text{ mol l}^{-1}$$

19 $n_{H_3PO_4} = C \times V = 0.1 \times 0.0167 = 0.00167$
From the equation 3 mol LiOH react with 1 mol of H_3PO_4.
Thus 0.00501 mol of LiOH react with 0.0167 mol of H_3PO_4.

$$C_{LiOH} = \frac{n}{V} = \frac{0.00501}{0.01} = 0.50 \text{ mol l}^{-1}$$

20 a) Sulphuric acid and nitric acid.
b) Sulphur dioxide and nitrogen dioxide.
c) i) Sulphur dioxide, ii) nitrogen dioxide.

21 a) $NH_3 + HNO_3 \rightarrow NH_4NO_3$
b) $2NH_3 + H_2SO_4 \rightarrow (NH_4)_2SO_4$
c) $KOH + HNO_3 \rightarrow KNO_3 + H_2O$
d) $3NH_3 + H_3PO_4 \rightarrow (NH_4)_3PO_4$

22 $n_{NaOH} = C \times V = 0.5 \times 0.0334 = 0.0167$
From equation $n_{NaOH} = n_{ethanoic\ acid} = 0.0167$

$$C_{ethanoic\ acid} = \frac{n}{V} = \frac{0.0167}{0.020} = 0.835 \text{ mol l}^{-1}$$

3.3 Metals

1 a) B and F, b) A and E, c) D, d) B and F

2 C; **3** A; **4** A

5 a) C, b) A, c) D, d) A

6 D; **7** D

8 a) B
b) i) $Mg(s) + Fe^{2+}(aq) \rightarrow Mg^{2+}(aq) + Fe(s)$
ii) $Mg(s) \rightarrow Mg^{2+}(aq) + 2e^-$ (oxidation)
$2e^- + Fe^{2+}(aq) \rightarrow Fe(s)$ (reduction)

9 a) Copper + silver(I) nitrate solution
$\rightarrow$ silver + copper(II) nitrate solution
OR copper + silver(I) ions $\rightarrow$ silver + copper(II) ions
b) The nitrate ion
c) $Cu(s) \rightarrow Cu^{2+}(aq) + 2e^-$ (oxidation)
$2e^- + 2Ag^+(aq) \rightarrow 2Ag(s)$ (reduction)

10 a) One in which electrons are lost.
b) One in which electrons are gained.
c) One in which reduction and oxidation take place (one in which there is both loss and gain of electrons).

11 a) Those above hydrogen.
b) $Zn(s) + 2H^+(aq) \rightarrow Zn^{2+}(aq) + H_2(g)$. This is a redox reaction.
c) $Zn(s) \rightarrow Zn^{2+}(aq) + 2e^-$ (oxidation)
$2e^- + 2H^+(aq) \rightarrow H_2(g)$ (reduction)

12 a) $Zn^{2+}(Cl^-)_2$
b) i) $Zn^{2+} + 2e^- \rightarrow Zn$ (reduction)
ii) $2Cl^- \rightarrow Cl_2 + 2e^-$ (oxidation)

13 a) Reduction.
b) From beaker A to beaker B.
c) $Zn(s) \rightarrow Zn^{2+}(aq) + 2e^-$

14 a) By observing how vigorously/rapidly they react.
b) Particle size.
c) i) $4Na + O_2 \rightarrow 2Na_2O$
ii) $2Ca + O_2 \rightarrow 2CaO$
iii) $4Al + 3O_2 \rightarrow 2Al_2O_3$

15 a) Li, Na, K, Rb, Cs, Fr.
b) Oxygen.
c) Lithium, sodium and potassium.
d) $2M + 2H_2O \rightarrow 2MOH + H_2$

16 ... unreactive ... silver ... reactive ... zinc ... carbon ... aluminium ... electrolysis

17 a) By incomplete combustion of carbon.
b) $Fe_2O_3 + 3CO \rightarrow 2Fe + 3CO_2$
c) To remove impurities.

18 a) The flow of ions through electrolytes accompanied by chemical changes at the electrodes which can result in the decomposition of the electrolyte. (Literal meaning – the breaking up of a compound by means of electricity.)
b) At the negative electrode:
$Na^+ + e^- \rightarrow Na$ (reduction)
At the positive electrode:
$2Cl^- \rightarrow Cl_2 + 2e^-$ (oxidation)
c) The ions are not free to move

19 a) A chemical reaction which involves the surface of the metal changing from an element to a compound.
b) Iron is more reactive than copper.
c) Rusting is only used to describe the corrosion of iron (or steel).
d) Water and oxygen.

20 a) i) $Fe^{2+}(aq)$
ii) $Fe(s) \rightarrow Fe^{2+}(aq) + 2e^-$
c) i) $OH^-(aq)$
ii) $4e^- + 2H_2O(l) + O_2(aq) \rightarrow 4OH^-(aq)$

21 a) Acids react with iron and act as electrolytes in the chemical cells which are set up on the surface of the metal during the corrosion.
b) When dissolved in water common salt acts as an electrolyte because it contains ions. Sugars do not produce ions when dissolved and therefore do not act as electrolytes.

22 The battery causes electrons to flow away from the iron nail attached to the positive electrode, which loses electrons and therefore corrodes. The $Fe^{2+}(aq)$ ions produced

$$Fe(s) \rightarrow Fe^{2+}(aq) + 2e^-$$

give a blue colour with ferroxyl indicator. The battery causes electrons to flow towards the iron nail attached to the negative electrode, which is thus prevented from losing electrons and does not corrode. These electrons are gained by oxygen and water producing $OH^-(aq)$ ions

$$4e^- + 2H_2O(l) + O_2(aq) \rightarrow 4OH^-(aq)$$

which give a pink colour with ferroxyl indicator.

23 a) By dipping in molten zinc.
b) Galvanising.
c) Because zinc is above iron in the ECS, electrons flow from the zinc to protect the iron.
d) Zinc loses electrons forming Zn^{2+} ions and is therefore sacrificed to protect the iron.

24 a) The layer of tin keeps out oxygen and water.
b) Iron is above tin in the ECS. When the tin coating is broken, electrons flow from the iron (which is sacrificed) to the tin (which is protected).

25 a) The negative electrode.
b) Any soluble zinc salt – for example zinc chloride, zinc sulphate or zinc nitrate.
$$Zn^{2+}(aq) + 2e^- \rightarrow Zn(s)$$

26 a) The electrons flow from electrode B to A through the wire.
b) $2Ag^+ + 2I^- \rightarrow 2Ag + I_2$
c) It completes the circuit.
d) Add starch solution and a deep blue/black colour is formed.

SECTION 4 WHOLE-COURSE QUESTIONS

4.1 Whole-course questions

1 a) 7, b) 3, c) 3, d) 4, e) 2, f) 2
g) 23, h) 12, i) 12, j) 11, k) 10, l) 2, 8
m) $^{27}_{13}Al^{3+}$, n) 13, o) 14, p) 10

2 a) 19, b) 9, c) 9, d) 10, e) 10, f) 2, 8
g) 34, h) 16, i) 16, j) 18, k) 18, l) 2, 8, 8
m) $^{14}_{7}N^{3-}$, n) 14, o) 7, p) 2, 8

3 a) i) The ions are not free to move.
ii) The ionic bonding is strong.
iii) $Ba^{2+} + 2e^- \rightarrow Ba$
b) i) $2Al + 3BaO \rightarrow Al_2O_3 + 3Ba$
ii) Redox (could also be considered to be a type of displacement).

4 a) i) $Mg^{2+}(aq) + 2OH^-(aq)$
$\rightarrow Mg^{2+}(OH^-)_2(s)$
ii) Filtration
b) $Mg(OH)_2 \rightarrow MgO + H_2O$
c) i) $MgO + C + Cl_2 \rightarrow MgCl_2 + CO$
ii) Hydrochloric acid
d) i) $Mg^{2+} + 2e^- \rightarrow Mg$ (reduction)
$2Cl^- \rightarrow Cl_2 + 2e^-$ (oxidation)
ii) The chlorine which is required for the production of magnesium chloride from magnesium oxide is provided by the electrolysis of the molten magnesium chloride.

5 a)

b) $CaCO_3 + 2HCl \rightarrow CaCl_2 + H_2O + CO_2$
c) i)

$$\text{Average rate} = \frac{\text{change in quantity monitored}}{\text{time taken}}$$

$$= \frac{2.7}{2} = 1.35 \text{ g min}^{-1}$$

ii) $\frac{4.1 - 2.7}{2} = \frac{1.4}{2} = 0.7 \text{ g min}^{-1}$

d) The acid concentration decreased (the surface area of the marble also decreased).
e) The concentration of the acid could have been higher; the temperature could have been higher; the marble chips could have been smaller.

6 a) The reaction is reversible.
b) Heterogeneous.
c) A catalyst is said to be poisoned when impurities are preferentially adsorbed onto

the surface of the catalyst, blocking active sites and thus reducing catalyst efficiency.

7 a) i) $2C + O_2 \rightarrow 2CO$
ii) $Fe_2O_3 + 3CO \rightarrow 2Fe + 3CO_2$
iii) $CaCO_3 + SiO_2 \rightarrow CaSiO_3 + CO_2$
b) i) Solid aluminium oxide does not conduct because its ions are not free to move.
ii) $Al^{3+} + 3e^- \rightarrow Al$
iii) Liquid
iv) The oxygen, which is released at this electrode, reacts with the hot carbon to form carbon dioxide and/or carbon monoxide.
c) Aluminium is higher than iron in the reactivity series and therefore 'holds on' to oxygen more strongly than iron.

8 a) $H - C \equiv C - H$
b) $2C_2H_2 + 5O_2 \rightarrow 4CO_2 + 2H_2O$
c) i) C_nH_{2n-2}
ii) The members have similar chemical properties and show a gradual change in physical properties.

9 a) $C_5H_{10}O$
b) $C_nH_{2n}O$
c) Hexan-2-one
d) Pentan-2-ol

10 a) Cyclobutane

$$
\begin{array}{c}
\text{H} \quad \text{H} \\
| \quad\; | \\
\text{H} - \text{C} - \text{C} - \text{H} \\
| \quad\; | \\
\text{H} - \text{C} - \text{C} - \text{H} \\
| \quad\; | \\
\text{H} \quad \text{H}
\end{array}
$$

b) Y

$$
\begin{array}{c}
\text{H} - \text{C} - \text{C} \overset{\text{H}}{\nearrow} \\
\quad| \quad\;\; \| \\
\text{H} - \text{C} - \text{C} \\
\quad| \quad\;\; \searrow \text{H} \\
\quad\text{H}
\end{array}
$$

Z

$$
\begin{array}{c}
\text{H} \quad \text{H} \\
| \quad\; | \\
\text{H} - \text{C} - \text{C} - \text{Br} \\
| \quad\; | \\
\text{H} - \text{C} - \text{C} - \text{Br} \\
| \quad\; | \\
\text{H} \quad \text{H}
\end{array}
$$

11 a) An addition reaction in which water adds on to a molecule in the presence of a catalyst.

b)

$$
\begin{array}{c}
\text{H} \quad \text{H} \quad\quad \text{H} \\
| \quad\; | \quad\quad\; \nearrow \\
\text{H} - \text{C} - \text{C} = \text{C} \\
| \quad\; | \quad\quad\; \searrow \\
\text{H} \quad \text{H} \quad\quad \text{H}
\end{array}
\quad + \quad \text{H} - \text{O} \diagdown \text{H}
$$

$$
\rightarrow \quad
\begin{array}{c}
\text{H} \quad \text{H} \quad \text{H} \\
| \quad\; | \quad\; | \\
\text{H} - \text{C} - \text{C} - \text{C} - \text{H} \\
| \quad\; | \quad\; | \\
\text{H} \quad \text{O} \quad \text{H} \\
\quad\quad | \\
\quad\quad \text{H}
\end{array}
$$

c) Propan-1-ol

$$
\begin{array}{c}
\text{H} \quad \text{H} \quad \text{H} \\
| \quad\; | \quad\; | \\
\text{H} - \text{C} - \text{C} - \text{C} - \text{H} \\
| \quad\; | \quad\; | \\
\text{H} \quad \text{H} \quad \text{O} \\
\quad\quad\quad\quad | \\
\quad\quad\quad\quad \text{H}
\end{array}
$$

12 a) Propan-1-ol and ethanoic acid
b) Concentrated sulphuric acid
c)

$$
\begin{array}{c}
\text{H} \quad\quad \text{O} \\
| \quad\quad\; \nearrow\!\!\!\!/ \\
\text{H} - \text{C} - \text{C} \\
| \quad\quad\; \searrow \\
\text{H} \quad\quad \text{O} - \text{H}
\end{array}
\quad + \quad
\begin{array}{c}
\text{H} \quad \text{H} \quad \text{H} \\
| \quad\; | \quad\; | \\
\text{H} - \text{O} - \text{C} - \text{C} - \text{C} - \text{H} \\
\quad\quad\quad | \quad\; | \quad\; | \\
\quad\quad\quad \text{H} \quad \text{H} \quad \text{H}
\end{array}
$$

$$
\rightleftharpoons
\begin{array}{c}
\text{H} \quad\quad \text{O} \\
| \quad\quad\; \nearrow\!\!\!\!/ \\
\text{H} - \text{C} - \text{C} \\
| \quad\quad\; \searrow \\
\text{H} \quad\quad \text{O} - \text{C} - \text{C} - \text{C} - \text{H}
\end{array}
\quad + \quad \text{H} - \text{O} \diagdown \text{H}
$$

d) Condensation

13 a) One which has been formed by the joining together of many monomer molecules to form the polymer and nothing else.
b) The hydroxyl group.
c) Four.
d)

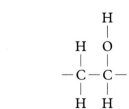

e)

<pre>
              H
              |
      H   O
      | |
      C = C
      | |
      H   H
</pre>

14 a)

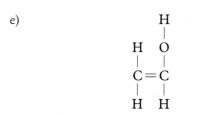

b)

<pre>
      O   H
      || |
   -- C — N --
</pre>

c)

<pre>
        H / H \ O
        | | | ||
  H — N + C + C — O — H
        | | |
        \ H / 5
</pre>

d) Condensation.

15 a) It is not glucose, fructose or maltose.
b) It is not starch.
c) Hydrolysis.
d) Blue to orange-red.
e) Glucose and fructose, $C_6H_{12}O_6$
f) $C_{12}H_{22}O_{11} + H_2O \rightarrow C_6H_{12}O_6 + C_6H_{12}O_6$
g) Sucrose.

16 a)

<pre>
      H   H   O
      | | ||
  H — N — C — C — O — H
          |
          H — C — H
          |
          H

      H   H   O
      | | ||
  H — N — C — C — O — H
          |
          H — C — H
          |
          H — C — H
          |
          S
          |
          H — C — H
          |
          H
</pre>

b) Homogeneous.

c)

<pre>
  H H O H H O H H O
  | | || | | || | | ||
H—N—C—C—N—C—C—N—C—C—O—H
     | | |
     H—C—H   H—C—H   H
     | |
     H   H—C—H
              |
              S
              |
              H—C—H
              |
              H
</pre>

d) The amine (or amino) group

<pre>
        H
       /
  — N
       \
        H
</pre>

and the carboxyl group

<pre>
        O
       //
  — C
       \
        O — H
</pre>

17 a) Energy.
b) i) Esters.
ii)

<pre>
        H
        |
  H — C — O — H
        |
  H — C — O — H
        |
  H — C — O — H
        |
        H
</pre>

iii) 3
c) Hydrogen molecules add on to C = C bonds in the oils, removing the unsaturation and producing C — C bonds.
... $- CH = CH - + H_2 \rightarrow - CH_2 - CH_2 -$

18 a) $2Ca + O_2 \rightarrow 2CaO$
$CaO + H_2O \rightarrow Ca(OH)_2$
b) $n_{Ca(OH)_2} = \dfrac{m}{fm} = \dfrac{2}{74} = 0.027$

$C_{Ca(OH)_2} = \dfrac{n}{V} = \dfrac{0.027}{0.5} = 0.54 \text{ mol l}^{-1}$

c) It is completely dissociated into ions in aqueous solution.

19 a) Hydrochloric acid.
b) Ethanoic acid is a weak acid dissociating only partly in aqueous solution to produce

hydrogen ions and ethanoate ions. Hydrochloric acid is a strong acid, dissociating completely to produce hydrogen ions and chloride ions. Since the 0.1 mol l^{-1} hydrochloric acid has the higher concentration of hydrogen ions it has the lower pH value.

   c) $CH_3COOH(aq) \rightleftharpoons CH_3COO^-(aq) + H^+(aq)$

20 a) i) $Fe_2O_3 + 6HCl \rightarrow 2FeCl_3 + 3H_2O$

     ii) $(Fe^{3+})_2(O^{2-})_3(s) + 6H^+(aq) + 6Cl^-(aq) \rightarrow 2Fe^{3+}(aq) + 6Cl^-(aq) + 3H_2O(l)$

   b) $n_{NaOH} = C \times V = 0.2 = 0.01 = 0.002$

From the equation, 1 mol of NaOH reacts with 1 mol of HCl.

So 0.002 mol of NaOH reacts with 0.002 mol of HCl.

$$C_{HCl} = \frac{n}{V} = \frac{0.002}{0.0092} = 0.217 \text{ mol l}^{-1}$$

But this is the diluted solution, so the concentration of the original hydrochloric acid = $0.217 \times 50 = 10.85$ mol l^{-1}

21 a) Carbon dioxide.

   b) Sodium sulphate.

   c) i) $Na_2CO_3 + H_2SO_4 \rightarrow Na_2SO_4 + H_2O + CO_2$

     ii) $n_{Na_2CO_3} = C \times V = 0.12 \times 0.02 = 0.0024$

From the equation, 1 mol of Na_2CO_3 reacts with 1 mol of H_2SO_4.

So 0.0024 mol of Na_2CO_3 reacts with 0.0024 mol of H_2SO_4.

$$V_{H_2SO_4} = \frac{n}{C} = \frac{0.0024}{0.1} = 0.024 \text{ litres}$$

$$= 24 \text{ cm}^3$$

22 a) From zinc to copper.

   b) $Zn(s) \rightarrow Zn^{2+}(aq) + 2e^-$ (oxidation)
$2e^- + Cu^{2+}(aq) \rightarrow Cu(s)$ (reduction)

   c) Copper.

   d) The voltage would increase.

23 a) i) The hydroxide ion, $OH^-(aq)$

     ii) The iron(II) ion, $Fe^{2+}(aq)$

   b) Magnesium is above iron in the ECS and therefore electrons flow from the magnesium, which is sacrificed, to the iron which is protected (and does not lose electrons forming $Fe^{2+}(aq)$ ions).

   c) i) Tin is a fairly unreactive metal and provides good physical protection for iron, while intact, keeping out oxygen and water.

     ii) Iron is above tin in the ECS and therefore when the tin coating is broken electrons flow from the iron, which is sacrificed and corrodes, to the tin which is protected.
$O_2 + 2H_2O + 4e^- \rightarrow 4OH^-$

24 a) i) E, ii) C, iii) F, iv) A, v) B

   b) Combustion/burning

25 a) Addition polymerisation

   b) (Catalytic) hydration/addition

   c) Hydrolysis

   d) Dehydration

   e) Displacement/redox

   f) Fermentation

   g) Reduction

   h) Condensation polymerisation

   i) Precipitation

   j) Oxidation

   k) Decomposition

4.2 Questions involving Prescribed Practical Activities

1 a) From colourless to dark blue.

   b) i) 0.0148 s^{-1}

     ii) $\frac{1}{t} = 0.0148$ s^{-1},

so $t = \frac{1}{0.0148} = 67.6$ s

   c) Reaction rate is directly proportional to the sodium persulphate concentration.

   d) • The temperature of the reaction mixture.
     • The volume and concentration of the solution containing potassium iodide and sodium thiosulphate (effectively the masses of potassium iodide and sodium thiosulphate).

2 a) A cross is lightly drawn on a piece of paper and a small beaker is placed on this. A measured volume of sodium thiosulphate solution is added to the beaker which is followed by a measured volume of dilute hydrochloric acid, at which point timing is started and the mixture stirred. Timing is stopped when the cross, viewed from above, disappears.

b)

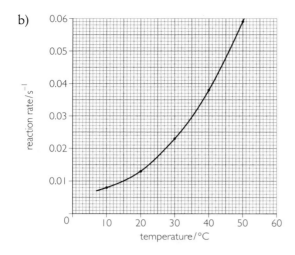

reaction rate/s^{-1} (vertical axis)
temperature/°C (horizontal axis)

c) $\frac{1}{t} = 0.059 \text{ s}^{-1}$, so $t = \frac{1}{0.059} = 17$ s

d) • The volume and concentration of the hydrochloric acid.
   • The volume and concentration of the sodium thiosulphate solution.

3 a)

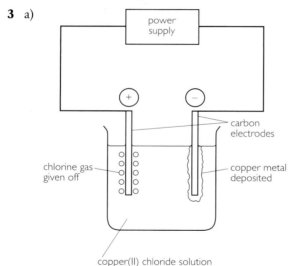

power supply

+   −

carbon electrodes

chlorine gas given off

copper metal deposited

copper(II) chloride solution

b) A brown solid forms on the electrode.
c) i) Chlorine gas.
   ii) Moist pH paper (or blue litmus paper) held just above the electrode first of all turns red and is then bleached.
d) Chlorine gas is toxic/poisonous. Only a small amount should be produced/sniffed. Alternatively use a fume cupboard.

4 a) Indicates the presence of a carbon to carbon double covalent bond, C = C.

b) Shake in a stoppered test tube with bromine solution. Decolourisation indicates unsaturation.
c) But-l-ene

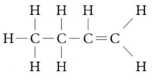

But-2-ene

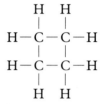

d) Cyclobutane

e) Bromine solution is toxic/poisonous. Carry out experiment in a fume cupboard.

5 a) X = ceramic wool
      Y = aluminium oxide
      Z = bromine solution.
   b) The breaking up of large hydrocarbon molecules to produce smaller molecules.
   c) To allow the reaction to take place at a lower temperature.
   d) Red/orange/yellow to colourless.
   e) As the test tube cools, the pressure inside it drops and air pressure forces the liquid up the tube – i.e. 'suck back' takes place.
   f) Decane, $C_{10}H_{22}$
   g) There are insufficient hydrogen atoms.

6 a) Starch produces a dark blue colour with iodine solution.
   b) One in which a large molecule is broken down into two or more smaller ones by reaction with water.
   c) Add Benedict's solution and heat. The colour change would be blue to orange/red.
   d) Amylase.
   e) To show that water alone does not hydrolyse starch under the experimental conditions used.

7 a) Magnesium sulphate, water and carbon dioxide.
   b) Add magnesium carbonate to the dilute sulphuric acid until the acid is completely

148

neutralised, as indicated by no more carbon dioxide being given off, the presence of unreacted solid magnesium carbonate and pH paper indicating a neutral solution (green). The excess magnesium carbonate is filtered off and some of the water evaporated from the filtrate, which is then set aside to crystallise.

c) Hydrogen gas is given off which is flammable – therefore no naked flames.

8 a) The further apart the two metals are in the ECS the greater is the voltage produced.
b) Aluminium oxide.
c) Any three from:
- temperature of the electrolyte solution
- same electrolyte
- same volume and concentration of electrolyte solution
- same size, distance apart and depth of electrodes.

9 a) Potassium permanganate.
b) The metals do not react with oxygen when cold.
c) Glowed very brightly, changed from silvery to white.
d) Same particle size and mass (or number of moles).

Appendix: Types of chemical formulae

1 a) 24, b) CH_2O

2 SiC

3

$$H \diagdown_{\diagup} N \cdots_{\diagdown} ^{H}_{H}$$

4 $C_6H_8O_7$

5 a) i) C_2H_6 ii) CH_3 iii)

$$H-\overset{\overset{\displaystyle H}{|}}{\underset{\underset{\displaystyle H}{|}}{C}}-\overset{\overset{\displaystyle H}{|}}{\underset{\underset{\displaystyle H}{|}}{C}}-H$$

iv) CH_3CH_3

b) i) C_3H_6 ii) CH_2

iii)

$$H-\overset{\overset{\displaystyle H}{|}}{\underset{\underset{\displaystyle H}{|}}{C}}-\overset{\overset{\displaystyle H}{|}}{\underset{\underset{\displaystyle H}{|}}{C}}=C\diagup^{H}_{\diagdown H}$$

iv) $CH_3CH = CH_2$

c) i) C_4H_{10} ii) C_2H_5

iii)

$$H-\overset{\overset{\displaystyle H}{|}}{\underset{\underset{\displaystyle H}{|}}{C}}-\overset{\overset{\displaystyle H}{|}}{\underset{\underset{\displaystyle H}{|}}{C}}-\overset{\overset{\displaystyle H}{|}}{\underset{\underset{\displaystyle H}{|}}{C}}-\overset{\overset{\displaystyle H}{|}}{\underset{\underset{\displaystyle H}{|}}{C}}-H$$

iv) $CH_3CH_2 CH_2 CH_3$

d) i) C_4H_8 ii) CH_2

iii)

$$H-\overset{\overset{\displaystyle H}{|}}{\underset{\underset{\displaystyle H}{|}}{C}}-\overset{\overset{\displaystyle H}{|}}{C}=C-\overset{\overset{\displaystyle H}{|}}{\underset{\underset{\displaystyle H}{|}}{C}}-H$$

iv) $CH_3CH = CHCH_3$

e) i) C_6H_{12} ii) CH_2

iii)

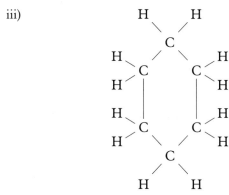

iv)

$$\begin{array}{c} CH_2 \\ \diagup \quad \diagdown \\ CH_2 \qquad CH_2 \\ | \qquad\qquad | \\ CH_2 \qquad CH_2 \\ \diagdown \quad \diagup \\ CH_2 \end{array}$$

6 $P_4 + 5O_2 \rightarrow P_4O_{10}$

7 a) $C_9H_8O_4$, b) $C_8H_9NO_2$

8 a) 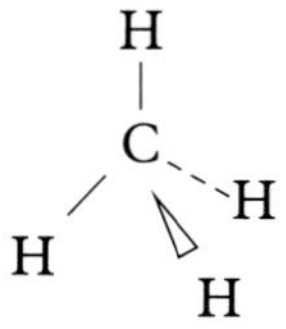   tetrahedral

b) 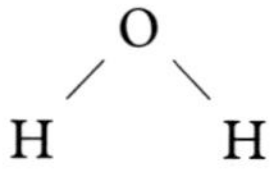   bent/angular

c) 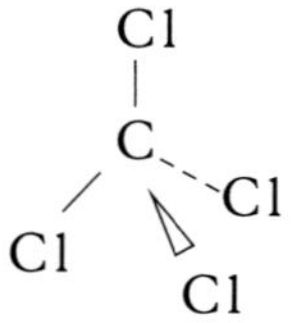   Cl   tetrahedral

      |
    C.
  / ⟍ ⟍Cl
Cl ⟍
    Cl

d) O = C = O   linear